Andreas Riener

Sensor-Actuator Supported Implicit Interaction in Driver Assistance Systems

VIEWEG+TEUBNER RESEARCH

Andreas Riener

Sensor-Actuator Supported Implicit Interaction in Driver Assistance Systems

With forewords by Univ.-Prof. Dr. Alois Ferscha and Prof. Dr. Albrecht Schmidt

VIEWEG+TEUBNER RESEARCH

Bibliographic information published by the Deutsche Nationalbibliothek
The Deutsche Nationalbibliothek lists this publication in the Deutsche Nationalbibliografie; detailed bibliographic data are available in the Internet at http://dnb.d-nb.de.

Dissertation Universität Linz, 2009

Gedruckt mit Unterstützung des Bundesministeriums
für Wissenschaft und Forschung in Wien.

1st Edition 2010

Editorial Office: Ute Wrasmann | Anita Wilke

Vieweg+Teubner is part of the specialist publishing group Springer Science+Business Media.
www.viewegteubner.de

Cover design: KünkelLopka Medienentwicklung, Heidelberg
Printing company: STRAUSS GMBH, Mörlenbach
Printed on acid-free paper
Printed in Germany

ISBN 978-3-8348-0963-6

Foreword

Technological advances, miniaturization of embedded computing technology and wireless communication together with the evolution of global networks like the Internet have brought the vision of pervasive and ubiquitous computing to life: technology seamlessly woven into the "fabric of everyday life". Along with this development goes the need and challenge of interfaces supporting an intuitive, unobtrusive and distraction free interaction with such technology-rich environments. Considering the huge amount, and ever growing number of a vast manifold of heterogeneous, small, embedded or mobile devices shaping the "pervasive computing landscape", makes traditional, explicit and attention based styles of interaction appear hopeless. To make computing part of everyday life, the interfacing must go beyond traditional explicit interaction: pervasive computing system designs will have to – and are already successfully attempting to – revert the principle of the user being in an active and attentive role, to one where technology is attentive and active. Interaction is becoming implicit.

Implicit interaction is based on two main concepts: perception and interpretation. Perception concerns gathering information about the environment and situations, usually involving (technological) sensors. Interpretation is the mechanism to understand the sensed data. Conceptually, perception and interpretation when combined are described as situational context. A system aware of its situational context does not have to be explicitly forced to act (by the user), but by collecting and interpreting information about its environment can autonomously trigger actions. Input is not necessarily explicitly stated or intentionally given, but the system understands the information it collects as input. The active, driving role in the interaction is thus moved from the user to the system. Consequently, the user does not have to be attentive and responsive to a plethora of devices, but the devices – as single entities or as ensembles of cooperative entities – develop a "sense" for the user, and act accordingly.

Andreas Riener has early identified driver-vehicle interfaces as one of these emerging "pervasive computing landscapes", i.e. complex configurations of technological system components and services, and "implicit interaction" therein as a major research challenge. Seen from the technological perspective, advanced driver assistance systems and in-vehicle information and entertainment systems have experienced an explosive growth over the past two decades, and are now among the most researched pervasive computing systems at the confluence of industrial and academic research endeavors. The search for knowledge from a computer science and engineering viewpoint is often motivated by practical as well as industrial applications and is often not far removed from the goal of practical utility. The spectrum of disciplines increasingly engaging to answer the research questions raised out of this domain is, however, overwhelming. Psychological and psycho-physiological research attempts to understand the multifaceted relations among the different modalities of stimulus and perception or sensation, the individual role and synesthesia of sense organs for perception, the modeling and recognition of vital state from biosignals, memorizing and recalling sensation, etc. Neurophysiological research and cognitive science addresses the issues of attention, memory, learning, perception and action from a more behavioral aspect. Even sociological research engages to explain and deliver empirical evidence for phenomena observed when socio-technical systems like driver-vehicle systems are used at large scale. The questions posed and the methods used to conduct research within this field are as varied as are the disciplines devoted to it.

The questions posed by Andreas Riener in this book concern in-vehicle interaction design issues, particularly the potential utility of haptics and vibrotactile stimulation, as well as driver and driver vital state recognition from embedded sensors, both from a technological perspective. Usability engineering research methods are adopted, and empirical evidence is delivered from cases involving real users and real systems.

Since the beginning of research in automotive user interfaces, a variety of scientific disciplines and research efforts have addressed the role of the sense of touch for perception, and passive interaction for articulation. The particular results, the individual findings, the related body of literature, however, have not yet led to a comprehensive theory of "haptic perception", or "implicit interaction". This work, for the first time, systematically sorts out previous contributions to answer interface design issues where haptic perception melts with implicit interaction in a technology-rich setting of everyday use: the car. It clarifies many facets of where the use of haptics is beneficial in driver-vehicle interactions, and exemplifies how current automotive interfaces can be improved. Manifold are the findings that concern the interaction loop from the driver to the vehicle and back, some of them elucidating basic principles of haptics in interface design - not restricted to the automotive domain. Manifold are also the findings that postulate human attention and situation awareness as the primary automotive interface design concerns. Fortunately, however, also many questions remain unanswered. Among the most fundamental ones that this book raises, is the scientific clarification of the process from human haptic per-ception to cognition, and models for memorizing, remembering, learning and forgetting - within and outside the domain of automotive interface applications.

Alois Ferscha

Foreword

I had the pleasure to be the external examiner for the PhD dissertation by Andreas Riener. At the Mensch and Computer conference in Lübeck, Germany, Andreas gave me a first draft. At first, I was a bit shocked by the size of the document (if I remember right – close to 300 pages). When I started to read the thesis I was pretty impressed by the amount of related work Andreas cited that I did not know before. For me this is always a good indicator that students have dived deep into their topic. The experimental part of the work is exciting in a similar way. I have come across a number of new approaches for experiments; my favorite one is where Alois Ferscha's passion for car racing becomes clearly evident. A further reason why his dissertation is exciting is that Andreas did the research in the context of challenging real-world projects and applications.

As the basic technologies (processing, wireless networks, sensors, and actuators) become more widely available they transform vehicles and in particular cars into interactive computer systems. Cars are currently becoming a platform and interaction in this domain is very challenging due to the primary task of driving. Embedding computing, communication, and services into vehicles is technically possible; however, making applications usable and safe to use while driving is still an open issue. This book describes research at the crossroads of pervasive computing and human computer interaction (HCI), addressing the question how to exploit explicit as well as implicit interaction for automotive user interfaces. The central part of the research is the assessment of using the human sense of touch and in particular tactile modalities as an additional interaction channel for drivers. How to efficiently sense the user's state and intentions and how to effectively communicate information to the user without increasing the cognitive load is a hard scientific challenge. With his prototypes, experiments, and empirical data Andreas Riener makes important contributions to this research field. With his work he moves forward and shows a systematic approach for understanding vibro-tactile interaction in the car. The research questions faced in this new domain are not as clear as in established fields and finding an appropriate and accepted research methodology is part of the work. Especially with an increase in complexity of functionality provided in the car creating user interfaces that are still easy and save to use is very difficult. The work presented in this book is practically applicable, relevant, and important.

The main contributions of the book are in two areas: (i) the assessment and use of sitting posture of the driver and (ii) the investigation of vibro-tactile communication and interaction as an additional channel in automotive HCI. Overall the research shows, based on empirical validation, that introducing these new modalities can lessen workload and improve reaction time. The research reported is anchored in a large number of prototypes and systems that Andreas Riener has developed during his research.

The body of work which is the basis of this book is remarkable. The work exemplarily explains a broad set of methods and tools and the results are relevant for designing novel car user interfaces. Despite my initial shock on the number of pages this work is worthwhile and very interesting to read.

Prof. Dr. Albrecht Schmidt

Preface

Advances in microelectronics and embedded systems technology have accelerated the evolution of Driver Assistance Systems towards more driving safety, comfort, entertainment, and wayfinding. The technological and qualitative progress, however, results in a steadily rising interaction complexity, information overload, and intricate interface designs. Assessing interaction designs for in-car assistance services is an emerging and vibrant field of research.

In this work, implicit interaction in Driver-Vehicle Interfaces is introduced for the purpose of relieving the driver of cognitive overload situations. Today's most frequently used sensory modalities *vision* (seeing) and *hearing* are very often highly charged due to a larger number and complexity of (*i*) Advanced Driver Assistance Systems (ADAS) and/or In-Vehicle Information Systems (IVIS), (*ii*) infotainment devices, (*iii*) communication appliances, and others.

These systems have originally been developed for providing support to the vehicle driver, but unfortunately their operation and monitoring increasingly have the disadvantage of resulting in a high cognitive load for the driving person and consequently lead to operation errors, caused by (*i*) overlooking information or (*ii*) missing important messages. Another problem category are modality-based distraction factors, such as glaring or reflecting light, fog, snowfall, day and night vision, and changing light conditions for the visual channel, or motor and traffic noise, car stereo, passenger communication, and cell phone calls for the auditory notification channel.

Apart from these communication restrictions, possible solutions for the above-mentioned problems have been investigated and three suitable options have been identified: (*i*) the introduction of one or more additional feedback channels (a feedback channel in this context is a dedicated way for the driver to interact with the vehicle), (*ii*) the optimization of existing feedback modalities (e. g., consider age or sex-related differences for the visual, auditory, and other interaction channels), and (*iii*) the application of multimodal interfaces (this is a simultaneous activation or stimulation of two or more feedback channels for the purpose of notifying about only one informational item) in contrast to the primarily applied unimodal ones.

The introduction and utilization of multimodality is straightforward and, therefore, has been excluded from this work; the remaining two options have been extensively analyzed with both simulated and on-the-road experiments.

This research work is structured as follows: The introductory Part I defines commonly used terms, describes traditional concepts of information processing between a driver and the vehicle, indicates problems occurring in modern cars, and concludes with a definition of hypotheses and research questions. Part II, which is dedicated to an investigation of Driver-2-Vehicle (D2V) interfaces, summarizes related work on vibro-tactile stimulation with the aim to motivate the utilization of the sense of touch. Part III considers interaction from a driver's view, in particular its notification demands as well as possible causes of distraction, and gives suggestions for future Advanced Driver Assistance Systems incorporating vibro-tactile input and output. The methodical Part IV starts with an introduction of the utilized analytical methods, followed by a detailed description of the conducted experiments including results and considerations for their usage in novel applications and system designs for assisting implicit driver-vehicle input or car-driver output. Part V summarizes the findings of the experiments, annotates lessons learned during the experimental conduction, and finally concludes this work with some propositions for future research. The Appendix gives a description of utilized hardware components, discusses the basics of biometric identification and alphabets related to touch, and tries to clarify the usage of additional sensory modalities such as proprioception, temperature, and pain.

Acknowledgements

Working in a specific research area and especially on a dissertation (the basis of this book) is a highly interesting and important period in a person's career. However, the work is often exhausting and one is occasionally plagued by doubts about whether the task can ever be successfully completed.

A lot of persons are responsible for the success or failure of such a project. In my case, I would like to thank the following individuals who made substantial contributions to this work in different ways. Without their help it would not have been possible to complete both the doctoral dissertation and this book in such quality and detail.

- *Alois Ferscha*, my dissertation advisor, for his excellent mentoring and supervision throughout the entire process. He gave me the opportunity to carry out research in one of the most challenging areas of human-computer interaction (HCI) – the automotive field –, and furthermore encouraged me in difficult situations and times of uncertainty.

- *Albrecht Schmidt*, who was my second dissertation advisor. He always found time for discussions and made constructive suggestions for improvements. Furthermore, Professor Schmidt was responsible for the important task of considering the research challenges from a different angle, detached from my own narrow perspective.

- *Gabriel Robles-de-la-Torre*, *Susan Lederman*, *Roberta Klatzky*, *Vincent Hayward*, *Hong Z. Tan*, and *Roger Cholewiak*, all scientists and/or psychologists, for their attempts to clarify the ambiguously used terms related to the sense of touch.

- All my colleagues at work for their useful comments and their willingness to discuss various points. Cordial thanks are expressed to the following persons: *Bernadette Emsenhuber* for providing me with scribbles, sketches, and other graphic materials, often just in time, *Dominik Hochreiter* for his prolific knowledge and technical assistance in the implementation of analysis software and for his assistance in experimental processing, *Michael Matscheko* for the establishment of communication interfaces and analysis routines in the early stages of this work, and *Martin Mascherbauer* and *Martin Weniger*, who did an internship under my supervision, for their help with the realization of trace-driven and on-the-road experiments and associated interface adaptations. Their commitment resulted in improved experimental prototypes, finally leading to the results presented in the experimental section of my dissertation and this book.

- *Bernhard Schenkenfelder* and *Anthony Rastrick* for voluntarily proofreading the preliminary version of my dissertation. They provided valuable comments not only on typos and syntax errors, but also concerning the contents. *Andrea Pellette* was responsible for the editorial proofreading of the final version of this book – many thanks to her for this very hard but important work.

- *Anita Wilke*, editor of Vieweg+Teubner, Wiesbaden, who is responsible for the publication of this monograph, for her excellent cooperation throughout the entire publication process.

- My partner *Eva-Maria Höffinger*. She supported me with lots of patience and encouragement during the last two years of my PhD and during the editing of this book. She is the reason I did not lose courage in difficult times. Further thanks are owed to my parents and brothers for their unstinting support; they sustained me while I was completing my PhD – which, of course, was not always easy...

For all of the submitted conference papers and research articles, I would like to thank the co-authors and reviewers for their suggestions for improvements and their excellent cooperation. Their participation and the highly productive discussions have undoubtedly enhanced the quality of this book.

Andreas Riener

Contents

I Research Hypotheses 1

1 Introduction . . . 3
- 1.1 User Interfaces . . . 3
- 1.2 Interface Paradigms . . . 6
- 1.3 Distraction Forecast . . . 12

2 Perception . . . 14
- 2.1 Vision . . . 15
- 2.2 Hearing . . . 15
- 2.3 Touch . . . 16
- Summary and Impact . . . 17

3 Driver Expression . . . 18
- 3.1 Modes of Interaction: Explicit versus Implicit Feedback . . . 19
- 3.2 Modes of Interaction: Cardinality of Sensory Channels . . . 21
- Summary and Impact . . . 26

4 Perception and Articulation . . . 27
- 4.1 Application Domains . . . 27
- 4.2 Participation Demand . . . 28
- 4.3 Interaction Modalities versus Application Domains . . . 29

5 Hypotheses and Research Questions . . . 30
- 5.1 Interaction Modalities . . . 30
- 5.2 Research Questions . . . 31
- 5.3 Experiments . . . 32
- 5.4 Hypothesis I . . . 34
- 5.5 Hypothesis II . . . 35

Outline . . . 37

II Driver-Vehicle Interaction 39

6 Vibro-Tactile Articulation and Presentation . . . 41
- 6.1 Excursus: Sensory Modalities . . . 41
- 6.2 Motivation for Tactile Displays . . . 46

6.3 Definition of Terms 48
6.4 The Skin as a Sensory Organ 50
6.5 Research on Vibro-Tactile Stimulation 51
6.6 Touch-Based Driver-Vehicle Input 58
6.7 Haptic Interaction in Vehicles 58

III Information Needs of Drivers **65**

7 The Driver as the Weak Point in Interaction 67
7.1 Cognitive Load: A Big Challenge 67
7.2 Empirical Evidence for Cognitive Load 68
7.3 Managing Workload 69
8 Driver Activity and Notification Demands 71
8.1 Notification-Induced Driver Distraction 72
8.2 What Information? 74
8.3 When and How? 76
8.4 Where? 80
8.5 Visual and Auditory Perception 80
9 Advanced Driver Assistance Systems (ADAS) 83
9.1 Alternatives Supporting the Driver 84
10 Vibro-Tactile Interfaces 85
10.1 Motivation 86
10.2 Types of Stimuli 87
10.3 Stimulation via the Skin 89
10.4 Alphabets and Vibro-Tactile Patterns 93
10.5 Tactograms 98

IV Methodology **105**

11 Analytical Methods 107
11.1 Requirements and Technological Conditions 107
11.2 System Design 109
11.3 Eligible Methods for Pressure Sensing 110
11.4 Techniques for Vibro-Tactile Stimulation 113
12 Experiments 119
12.1 Identification and Authorization 121
12.2 Activity Recognition 132
12.3 Dynamic Adaptation of Vibro-Tactile Feedback 148

12.4 Simulating Real-Driving Performance . . . 161
12.5 Further Experiments . . . 182

V Discussion and Conclusion **183**

13 Predeterminations for Investigation . . . 185
13.1 Domain . . . 185
13.2 Prototype . . . 185
14 Reflecting on the Hypotheses . . . 186
14.1 On Implicit Driver Articulation . . . 186
14.2 On Vibro Tactile Driver Notification . . . 189
15 Experiments: Lessons Learned . . . 192
15.1 Novelty of Haptics . . . 192
15.2 Run-in Experiments . . . 192
15.3 Annoyances . . . 192
15.4 Varying Stimulation Parameters . . . 193
15.5 Parameter Mapping . . . 193
16 Conclusion . . . 194
16.1 Applicability . . . 194
16.2 Additional Workload for Touch Sensations? . . . 194
16.3 Limitations . . . 194
17 Future Prospects . . . 195
17.1 Reconfiguration of Vibro-Tactile Feedback Based on the Driver's Sitting Posture . . . 195
17.2 Reaction Times in Multimodal Interfaces . . . 195
17.3 Integration of Biochemical Features . . . 195
17.4 Additional Sensory Modalities . . . 196
17.5 Theoretical Studies on Haptics and Tactograms . . . 196

Appendices **199**

Bibliography **233**

Index **283**

List of Figures

1.1 Interaction modalities in vehicles (closed feedback loop). 5
1.2 Increased number of vehicle kilometers travelled in the US annually. 8
1.3 Increase in the number of cars (incl. cabs and publicly owned cars). 8
2.1 Perception times of individual feedback channels have to be aligned to one another in order to get meaningful results. 16
3.1 Appropriateness of combining modalities as redundant information sources (adapted from European Telecommunications Standards Institute (ETSI) [1]). 24
3.2 The amodal specification assumes that a given aspect of reality redundantly stimulates different channels of sense (adapted from Stoffregen *et al.* [2]). . . 25
4.1 Overview of vehicle-driver (output) and driver-vehicle (input) interaction modalities versus application domains with highlighted regions of research focus. 28
5.1 Experiments conducted for investigating driver expression and perception. . . 32
5.2 Driver-Vehicle Interaction (DVI) modalities versus application domains. . . . 33
6.1 Advantages and drawbacks of using the tactile sensory channel for interaction in vehicles (adapted from Erp [3, p. 23]). 48
6.2 Classification of the sense of touch. 49
6.3 Information flow in vehicular interfaces using touch sensation and manipulation (adapted from Tan [4]). 50
8.1 Driver activities and notification demands when operating a car. 71
8.2 The five *Allen relationships* applied to the four combination aspects temporal, spatial, syntactic, and semantic (adapted from Allen [5]). 72
8.3 Luminance density in road tunnels. 74
8.4 Categorization of driver information in vehicles into three classes (adapted from Kantowitz and Moyer [6, p. 3]). 75
8.5 Examples of suggested and not suggested auditory message length (adapted from Campbell *et al.* [7]). 78
9.1 Aspects of driver risks (adapted from Lerner *et al.* [8]). 83
9.2 Overview of Advanced Driver Assistance Systems (ADAS) (adapted from Bishop [9], van der Heijden *et al.* [10]). 85

9.3 Vital context in cars (adapted from Ferscha *et al.* [11] and extended by respiration frequency sensor, contactless ECG, pedal position sensors, and vibration elements). 86
10.1 Functional features of the four cutaneous mechanoreceptors, relevant for pressure/vibration sensation (the dashed frame indicates the type of receptor used). 90
10.2 Results of two-point discrimination threshold experiments, carried out by Kandal *et al.* [12], Gallace *et al.* [13], and Gibson [14], [15]. 93
10.3 Comparison of different alphabets (alternative forms of input) and their representation of typical characters, words, or activities (intent). 95
10.4 A *tactogram* is defined unambiguously by a 3D tactor pattern representation. The variation parameters for a specific tactor element are activation time, vibration intensity, and vibration frequency. 99
10.5 Static view of a multi-tactor system defining *tactograms* for three activities. . 100
10.6 The graph on the right side shows the dynamic flow of tactor activation for a multi-tactor system consisting of 2 x 5 elements. The tactile *image* is called a *tactogram* (or *vibration carpet*); individual elements are specified with a 6-tuple.. 101
10.7 Specific vibro-tactile patterns for controlling the user's *level of attention (LOA)*. 102
11.1 Threshold versus intensity (TVI) plot of Weber's law. 115
11.2 Threshold-intensity plot for the Weber-Fechner law. 115
11.3 TVI plot of Steven's formula for different sensory modalities. 116
11.4 Response behavior of Pacinian mechanoreceptors (PC) in relation to stimuli frequency. 117
12.1 Driver identification from sitting postures and its utilization potential in personalized vehicular services. 122
12.2 The prototype for driver identification, installed in a utility car, a sports car and a comfort station wagon (from left to right). 122
12.3 Illustration and localization of parameters used for the feature vector calculation. 123
12.4 Instructions for calculating the ranks (congruence classes). 128
12.5 Matching ranks for two subjects (31 and 105 samples). 130
12.6 The "Human Perception – Action" loop. 133
12.7 Activities, activity classes and corresponding sitting postures for an application in the vehicular domain. 134
12.8 The image on the left-hand side indicates a *soft* turn, the right picture shows a *hard* turn (both snapshots show a video of the test run together with the synchronized sitting posture). 136

12.9 On-the-road experiment for identifying correlations between body postures and steering activities. 137
12.10 Hypothesis on a correlation between body posture and cruising speed. 139
12.11 Fragmentation of the pressure-sensitive mat to determine the direction of leaning. 141
12.12 The GPS trace of one specific test run on the Wachauring Krems, Austria. . . 141
12.13 Driving studies for the experiment in activity recognition, conducted on the Wachauring Krems, Austria. 142
12.14 A driver's readiness to compensate for lateral acceleration forces in one test run (several laps). 144
12.15 Correlation of body posture to vehicle speed. 145
12.16 The visual analysis shows that the path of a body posture matches the curve of lateral acceleration. 146
12.17 The vibro-tactile car seat consisting of pressure-sensitive mats and arrays of tactile elements on both seat and back. 150
12.18 Mapping of feedback regions according to a driver's sitting position on the vibro-tactile seat. 152
12.19 Layout of the vibro-tactile feedback component for a vehicle seat (with respect to the relevant threshold distance). 153
12.20 The entire pressure sensing area on the seat mat, and the part currently covered by the driver (dashed frame). 154
12.21 Mat coverage on the seat. The planar plateau at height 34 is that region on the seat covered by all test candidates. 157
12.22 Contact with the back mat for 34 subjects. The plateau size is significantly smaller than that for the seat (see Fig. 12.21 on the left). 157
12.23 Mapping strategies for haptic feedback according to driver's sitting behavior. . 159
12.24 Haptic patterns for the activity *turn right* (from a driver's view), dynamically adapted to the region of the seat currently used by the driver (lightgray rectangles). 160
12.25 Perception times of individual feedback channels have to be aligned to each other in order to get meaningful results. 165
12.26 Valid task identifiers and their parameters. 166
12.27 Setup of the vibro-tactile seat used in the experiments and visual representation of two *tactograms* for right turns and switching the lights on/off. 167
12.28 Schematic representation of the experimental setting for trace-driven simulation in a garage. 168

12.29 The ATmega8-16PU microcontroller, placed on a STK500 development system, with external connectors and a voltage regulation circuit. 168
12.30 The garage with projection screen, data-acquisition vehicle, and processing equipment. 168
12.31 Experiment processing inside the car (note that the test participant is equipped with headphones). 168
12.32 Video frame with superimposed control window as shown in the projection (left), prototype of the vibro-tactile car seat (right). 169
12.33 Response times for auditory, visual and vibro-tactile stimuli (left column, from top) of male test persons only and a 5% confidence interval. The linear trend line on the response times runs downwards for all three notification modalities. The corresponding histograms in the right column show response times for the three notification modalities hearing, vision, and touch. Response to haptic notifications performs best, followed by visual and auditory sensations. 172
12.34 Boxplots show differences between male (left) and female (right) test participants. A salient contrast in minimum response time in favor of males can be indicated. 175
12.35 Boxplots for younger (left) compared to older (right) test participants. 176
12.36 The chart of mean response times for all modalities with superimposed linear trend line shows a slight decrease in response time (from 875*ms* to 855*ms*). 177
12.37 Correlation between response time and age, separately for the individual sensory modalities. 178
14.1 A specific driver's body posture (direction of leaning) correlated to the vehicle speed on a race course driven in counterclockwise direction. 187
14.2 The path of the body posture matches the curve of lateral acceleration (above a driver specific break-even speed). This effect can be used, for instance, for assistance systems operating proactively. 188
B.1 Comparison of biometrics (adapted from Liu *et al.* [16]). 209
B.2 Driver identification from sitting postures in vehicles (implementation as a universal, distributed pattern recognition system). 210
C.1 Physiological senses and corresponding fields of application in vehicles (adapted from Silbernagel [17]). 214
C.2 Interrelationship of kinesthesia (A), proprioception (B), and the vestibular sense (C) (taken from R. Cholewiak [18]). 221
C.3 Proprioception and kinesthesis, responsible for reception of stimuli produced within the body. 223

C.4 Proprioception refers to both kinesthetic and vestibular sensitivity (inspired by Cholewiak [18], [19]). 224
D.1 Model of system response time (adapted from Shneiderman *et al.* [20], Teal and Rudnicky [21]). 226
D.2 Midyear population by age and sex (Source: U.S. Census Bureau [22]). 229
E.1 Coding of the five vibration elements to 45 letters, digits, and words in the Vibratese language (adapted from Geldard [23]). 231

List of Tables

2.1 Humans' traditional and physiological senses. 14

12.1 Statistics on experimental data for the test on permanency. 126

12.2 The normalized confusion matrix of postures with artefacts for the seat (values in percent). 127

12.3 The normalized confusion matrix for the back mat (values in percent). 127

12.4 Accuracy of consecutive measurements of two subjects. 129

12.5 Tabular specification of features for dynamic activity recognition. 143

12.6 Threshold distances and resultant number of tactor elements for several parts of the body, relevant for in-seat sensing. 151

12.7 Statistics on mat coverage for the seat mat (population N=34, two thresholds 5%, 10%). Directions according to Fig. 12.20. 156

12.8 Mat coverage statistics for the back mat (population N=34, two thresholds 5%, 10%). 157

12.9 Evaluated activities and corresponding feedback signals. 165

12.10 Personal statistics of experiment participants, separated for males and females. 170

12.11 General statistics on response times for two confidence intervals and all male test persons, separated by modality. User response to haptic stimuli was best, far ahead of visual and auditory stimulations. 173

12.12 Gender-related statistics on response times, individually for the sensory channels vision, hearing, and touch (basic population of 792 data sets). Male test persons responded faster to stimuli than female participants (true for all three modalities). 174

12.13 Age-related statistics for male test persons only and separated by modality show almost no difference in the response times (basic population of 792 data sets). 175

List of Acronyms

Automotive Expressions

ABS Anti-Lock Braking System

ACC Adaptive Cruise Control

ADAS Advanced Driver Assistance Systems

AVC Active Vibration Control

BAS Brake Assist System

CAN Controller Area Network

CAS Collision Avoidance Systems

C2C Car-2-Car

C2I Car-2-Infrastructure

C2R Car-2-Roadside

DAS Driver Assistance Systems

D2C Driver-2-Car

D2V Driver-2-Vehicle

DOF Degrees Of Freedom

DVI Driver-Vehicle Interaction

EBA Emergency Brake Assist

ECU Electronic Control Unit

ESC Electronic Stability Control

FCW Forward Collision Warning

HVI Human-Vehicle Interaction

IDIS Intelligent Driver Information System

IMSIS In-Vehicle Motorist Services Information System

IRANS In-Vehicle Routing and Navigation System

ISIS In-Vehicle Signing Information System

IVIS In-Vehicle Information Systems

IVSAWS In-Vehicle Safety Advisory and Warning System

LDW Lane Departure Warning

LIN Local Interconnect Network

MOST Media Oriented Systems Transport

NHTSA National Highway Traffic Safety Administration

OBD On-Board Diagnostics

OLE Object Linking and Embedding

PDC Park Distance Control

RPM Revolution Per Minute

SRS Supplemental Restraint System

V2D Vehicle-2-Driver

WRAN Wireless Regional Area Network

Human-Computer Interaction, Medical, Others

ASL American Sign Language

ASR Automatic Speech Recognition

ECG Electrocardiogram

EDF European Data Format

GPS Global Positioning System

GSR Galvanic Skin Response

GVS Galvanic Vestibular Stimulation

HCI Human-Computer Interaction

HMI Human-Machine Interaction

HHSI Human Human System Interaction

HRT Human Reaction Time

IHCI Implicit Human-Computer Interaction

LOA Level of Attention

MAUI Minimal Attention User Interface

MRT Multiple Resource Theory

PC Pacinian Corpuscles

POI Point Of Interest

PUI Perceptual User Interface

RA Rapidly Adapting Afferent Neuron (Meissner Corpuscles)

SA1 Slowly Adapting Type 1 Afferent Neuron (Merkel Cells)

SA2 Slowly Adapting Type 2 Afferents (Ruffini Corpuscles)

SD Secure Digital

SLDS Spoken Language Dialog System

SWT Standard Widget Toolkit

TSAS Tactile Displays and Tactile Situation Awareness Systems

VPT Vibration Perception Threshold

WER Word Error Rate

WPM Words per Minute

Mathematical/Statistical Expressions

ANOVA Analysis of Variance

DBN Dynamic Belief Networks

FAR False Accept Rate

FRR False Reject Rate

HMM Hidden Markov Models

HHSMM Hierarchical Hidden Semi-Markov Models

IBL Instance-Based Learning

ICA Independent Component Analysis

JND Just Noticeable Difference

KLT Karhunen-Loeve Transformation

LDA Linear Discriminant Analysis

MDA Multivariate Data Analysis

MLR Multiple Linear Regression

PCA Principal Component Analysis

PCR Principal Component Regression

PDF Probability Density Function

PLS-R Partial Least Square Regression

SICMA Soft Independent Modeling of Class Analogy

SVD Single Value Decomposition

"The whole is more than the sum of its parts."
Aristotle (384 BC – 322 BC)
Greek critic, philosopher, physicist, and zoologist

Part I

Research Hypotheses

1 Introduction

1.1 User Interfaces

Human-Computer Interaction (HCI)[1] research has significantly changed over the last decades due to increasing complexity, technological advances, and successive embedding of tools and appliances in everyday life – the evolution of microprocessors, display devices, communication technologies and other electronic components has led to "innovative" Human-Computer Interfaces [25, p. 78]. The interfaces of the future will increasingly use gesture and speech recognition, intelligent agents, adaptive interfaces, etc. [26]. In conjunction, human performance and user experience with computer or information systems will remain an expanding research and development topic, as Shneiderman *et al.* [27, p. 4] indicated exemplarily.

Effective user interfaces, which are required to be designed for each field of application individually (see [28]), could offer humans more interaction comfort, or reduced cognitive load and distraction when incorporating users' actual requirements and desires.

1.1.1 Interaction in Vehicles

Traditional user interfaces for Human-Computer Interaction (as utilized in notebooks, PDAs, cell phones, or in operational controls for multimedia and home appliances) have attracted growing interest recently [27, p. 5], but concerns about Driver-Vehicle Interfaces meanwhile have been reported to a less amount although in-car electronics have also increased explosively.

As in many other fields of application, the trend in the automotive industry is towards electronic control systems. A few years ago, computer-assisted services and applications in a car focused on "non-critical tasks" such as comfort functions, but today even core vehicle functions are computer-controlled (e. g. engine control, brake assistance systems, adaptive cruise control, etc.) [29]. Reasons for this are (*i*) economically justified (the improved performance and reduced price of electronic devices forces the replacement of mechanical fallback systems by computer systems with active redundancy) or (*ii*) ecologically motivated (a computer chip or a piece of software is much lighter than for instance a mechanical brake system, and thus leads to a reduced gross weight of the vehicle and in succession to lower fuel consumption).

The latest analyses and studies have estimated that today's premium-class vehicles contain up to 70 application-specific Electronic Control Units (ECUs[2]), communicating over bus systems like CAN, FlexRay, LIN, or MOST; up to 90% of all innovations in the automotive domain (infotainment systems, Advanced Driver Assistance Systems, etc.) are provided or assisted by

[1]Definition: "Human-Computer Interaction is a discipline concerned with the design, evaluation and implementation of interactive computing systems for human use and with the study of major phenomena surrounding them" (working definition in the ACM SIGCHI Curricula for HCI, [24]).

[2]The abbreviation "ECU" is often, particularly in the automotive domain, referred to as "Engine Control Unit".

electronics or software [30], [31]. The future trend for information technology in the automotive domain can be deduced from the statement of Green [32, p. 844], who described (modern) vehicles as "JAVA-enabled browsers on wheels".

Assistance services and other in-car technology have the potential to improve driver and passenger safety as well as the driving comfort [33] on the one hand, however, they demand new styles of interaction (and therefore could become sources of distraction and increased cognitive load) on the other. The facilities introduced for assisting the driver in his/her operational tasks and relieving him from cognitive (over-)load are often the reason for new sources of distraction and additional risks of traffic accidents.

New means of interaction for vehicle operation have to be investigated in order to cope with these problems, while yet other appliances, functions and assistance systems will find their way into the vehicle [25] in the future.

1.1.2 Driver-Vehicle Interaction

Interfaces assisting the communication between a user and a system (and Driver-2-Vehicle (D2V) interfaces are a special peculiarity of these) ordinarily operate bilaterally: (*i*) feedback from an appliance or service is delivered to the user (this is referred to as "output") and (*ii*) any involved recipient responds on this stimulation with an adequate reaction (this direction of information flow is called "input"). A person involved in interaction either adapts his/her personal behavior, submits control actions toward the information transmitter or (re)adjusts the method of system handling.

When restricting the field of vision to the automotive domain, vehicle-control can be specified as a closed-loop[3], balanced feedback system: The driver provides input to the vehicle with designated control elements, such as steering wheel, gear-shift, or accelerator pedal, based on earlier obtained responses from the car and/or on the current traffic situation. Then he/she receives feedback from the corresponding vehicular components as an interpretation of his/her input (after processing), and subsequently reacts with back-transmission of a suitable "answer" to the system [34, p. 109].

More generally, the task of driving can be summarized as a complex cognitive task built up of the four sub-processes (*i*) perception, (*ii*) analysis, (*iii*) decision, and (*iv*) expression [35, p. 768], or slightly changed as "chain of sensory perception" [36, p. 3]. In this work only the two motoric sub-processes *perception* and *expression* are regarded; however, these two channels

[3] "Closed" in this context means, that there is a countable number of correct functions and failure situations, which can be fully covered in software or hardware (details on this topic are presented e. g. in [27, p. 78]). However, this holds true for a one-vehicle/one-driver system, but not for the nearly uncountable number of cars in dynamic road traffic environments – decisions in danger situations require negotiation among drivers, and are normally too complex to be automatically handled by the assistance systems.

of information flow (*input* for expression and *output* for perception, as indicated in Fig. 1.1) call for further qualification. Input is either given implicitly[4] or explicitly[5]; output is given using only a single sensory modality (*unimodal*) or by incorporating two or more channels of information simultaneously or sequentially (*multimodal*) [42, p. 3].

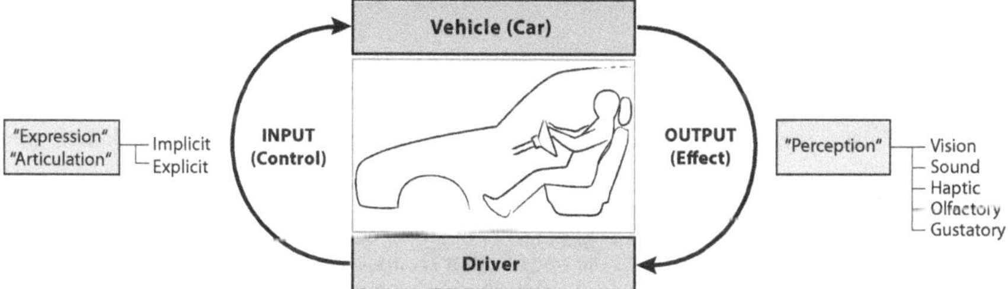

Fig. 1.1: Interaction modalities in vehicles (closed feedback loop).

A strong effort has been devoted to multimodal interfaces – they are believed to improve interaction performance. Nevertheless, a successful penetration of such systems requires adaptivity of the applied interaction channels depending on the field of application and a user's desires. For instance, a car driver wants to operate the navigation system by touch actions and visual output at a stop or by voice input and speech response when busy attending to traffic ([27, p. 347].) .

1.1.3 Single Driver, Single Vehicle

In this research work the profound field of Driver-Vehicle Interaction is restricted to considering only one specific aspect, namely systems comprising exactly one person (the driver; other passengers are excluded), and a single vehicle. For this reason, topics like car-to-car or vehicle-to-infrastructure communication can be disregarded – the focus is on an implicit interaction between a driver and his/her vehicle (either uni or multi-modal); explicit communication is no longer considered.

Related subtopics investigated in this work are for instance *Person Identification*, *Sitting Postures*, *Implicit Interaction*, *Interaction Modalities*, *Driving Comfort*, *Vehicle Handling*, *Car and Road Safety*, *Cognitive Load*, and *Distraction*.

[4]"Implicit" means that the system automatically detects user input at a certain level and that there is no active cooperation of the involved person necessary; of course, this requires the system's knowledge of behavior and/or context [37, p. 2].

[5]"Explicit" interaction is when the user is actively involved by expressing expected actions or activities towards the system; the system requires that the user attends to it to at least some degree; explicit interaction is believed to increase cognitive load, which could result in unwanted distractions, see [38, p. 2], [39], [40], or [41].

The remaining introduction section specifies the different factors influencing the Driver-Vehicle Interaction loop. Their potential will be pointed out with the objective of identifying opportunities for improving the communication between a driver and the vehicle (for instance, by reducing cognitive load using additional senses of perception).

1.2 Interface Paradigms

User interfaces in vehicles are different from their counterpart of traditional desktop interaction (commonly used input devices, such as fully functional keyboards, mice or trackballs, pen tablets, etc. are not available, output capabilities are limited due to fixed indicator elements in the dashboard, and screens are often not present or are too small). Beside these limitations it has been indicated, e. g. by Luk [43], that the one or other feedback option would not be appropriate in the vehicular context, e. g. information-rich displays would not be permitted while driving. Considering these issues, traditional interface paradigms have to be reworked or replaced by adequate alternatives to be suitable for in-car usage [44].

Apart from these technical restrictions, the development process of defining interfaces for the automotive domain has to consider a number of extraordinary issues, such as (*i*) the high dynamics of road traffic, (*ii*) changing conditions of illumination and ambient noise, (*iii*) the request for fail-safe and real-time operation (primary and secondary tasks contemporary demands for a driver's attention), etc.

All in all, vehicle user-interface design has to deal with:

(i) **Natural, intuitive interaction:** New modes of interaction have to be developed with the goal of being as intuitive[6] and natural[7] as possible and assuring the required level of safety at the same time. The development of safer Driver-Vehicle Interfaces without compromising the primary function of driving and concurrently considering the full range of operator behavior has become an ever important challenge in vehicle design.

(ii) **Sensory modalities:** The biggest part of information in vehicles is delivered with the modalities *vision* and *hearing* (approximately 80% of all sensory input is received via the eyes [25], [49], and another 15% via the auditory channel [50, p. 41].

[6]"Intuitive to use" technology is demanded because of the increasing ubiquity of interactive devices or applications. The time available for learning and using each resource decreases and coincidentally device complexity increases [45], [46]. Naumann *et al.* has defined the term *intuitive use* as "[..] a technical system is, in the context of a certain task, intuitively usable while the particular user is able to interact effectively, not-consciously using previous knowledge" [47].

[7]"Natural" human-machine interfaces require that the synergy of interaction be exploited through speech, body motion, gesture, etc. This can only be accomplished by a technology that allows for seamless, integrative and context-sensitive interpretation of information streams from different modalities [48, p. 25].

Wierda and Aasmann [51] formulated this fact concisely as "Driving is seeing [..]"). McCallum *et al.* [52, p. 32] reported, for example, an increase in driving performance, and a diminished cognitive workload of the driver when using speech interfaces instead of *normal* interaction paradigms for in-vehicle device control (but without considering the constraints of speech-interfaces).

Gender and age as well as the environmental influence of auditory and visual information channels[8] affect the response times for different driving situations and different groups of drivers. A possible way to compensate for these influences would be the application of additional sensory modalities, e. g. the sense of *touch*.

(iii) **Multimodality:** Driver-Vehicle Interaction could be improved by utilizing a combination of several sensory modalities for a specific stimulation. A number of studies (e. g. by Oviatt [56], Rachovides *et al.* [57], Vilimek and Zimmer [58], Wickens [59]) have confirmed that drivers (or humans in general) tend to interact multimodally.

(iv) **Interface rework:** Traditional interaction devices from desktop environments (such as keyboards, mice or large screens) are often inappropriate or unavailable in vehicles and have to be reworked or replaced by adequate alternatives [44].

(v) **Preventing miscommunication:** An interactive Driver-Vehicle System must be able to cope with the possibility of miscommunication[9]; the potential for problems is high, especially if the interaction occurs beyond the driver's attention or without his/her initiative (e. g. unintended actions or undesirable results) [38].

[8] There is evidence that voice is affected by a multitude of user-specific parameters like age, gender, driver's constitution, cognitive load or emotional stress; ambient noise and interference from conversations are furthermore responsible for distortion of spoken instructions [53, p. 368]; face or image detection often lacks illumination or suffers from background variation problems, slight changes in posture and/or illumination produce large changes in an object's visual representation [54] and in consequently results in performance loss for visual identification tasks [53, p. 21f], [55].

[9] "Miscommunication" includes misunderstanding (e. g. the driver obtains an interpretation that he/she believes is complete and correct, but the system meant another interpretation), non-understanding (a participant either fails to obtain any interpretation at all or obtains more than one interpretation without being able to choose among them), and misinterpretation (the most likely interpretation of a participant's utterance suggests that their beliefs about the world are unexpectedly out of alignment with the others) [60]. Misinterpretations and non-understandings are typically recognized immediately, whereas a participant is not aware if a misunderstanding occurs [61]. Successful communication requires that all the participants share considerable knowledge [62].

Over the last 30 years, the demands on vehicle drivers have increased dramatically due to technological advances and the steadily rising number of in-car information and communication systems [63].

Similarly, the role of the drivers has changed as well, since they are now often dealing with tasks of supervising and monitoring processes, performed automatically by in-car technology [64]. There is evidence that In-Vehicle Information and Communication Systems (IVIS), despite their obvious benefits, increase driver distraction and cognitive workload [65, p. 256] and may cause crashes or other safety problems [66], [64].

The following paragraphs are dedicated to an (incomplete) enumeration of distraction factors (explicitely perceived) – with the objective of identifying possibilities for reducing or resolving them later by using implicit technologies.

1.2.1 Surrounding Traffic

In the past fifty years, a steadily increasing volume of traffic, caused by a decreasing number of carpools (due to more flexible work schedules) and longer journeys to the place of work as well as the need of people for a more mobile lifestyle, has been evident (and this trend is expected to continue). The increasing number of cars on the roads (see Fig. 1.3) traveling longer distances (see Fig. 1.2) has led to a significant growth in cognitive workload for the driver [67] and is one of the determinants for distraction (which directly influences the rate of traffic accidents).

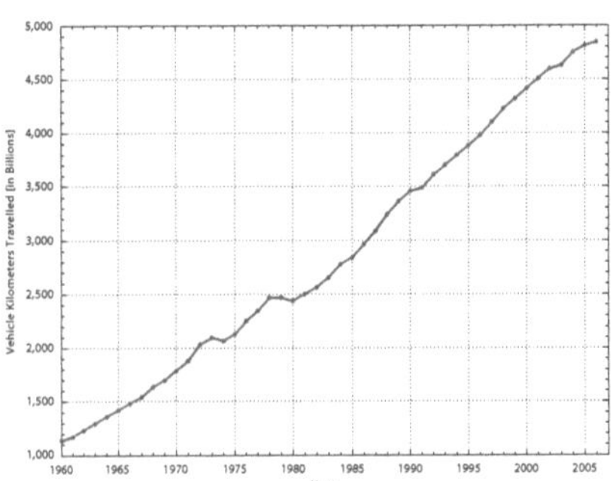

Fig. 1.2: Increased number of vehicle kilometers travelled in the US annually.

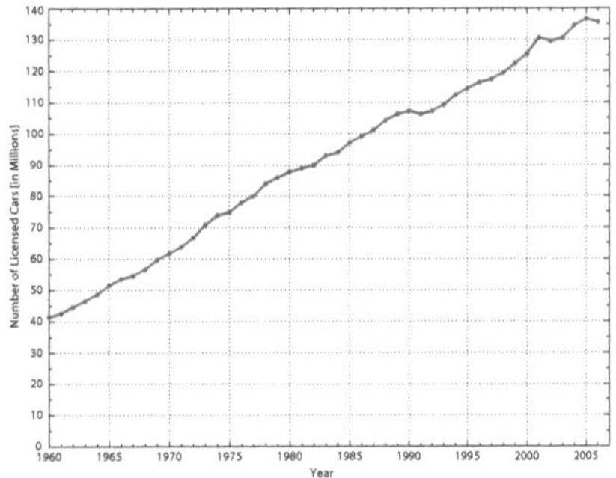

Fig. 1.3: Increase in the number of cars (incl. cabs and publicly owned cars).

A major challenge for a driver today is to keep his/her focus of attention on the primary driving task. Heavy (peak hour) traffic, a larger number of traffic signs, distractions inside the car (radio, noisy passengers, cell phones), overwhelming traffic regulations, etc. make it more and more difficult to meet this requirement [68]. Furthermore, Driver Assistance Systems will additionally make it difficult to attain this objective. Their number and communication demand increase as well, encouraging the distraction from the main driving task by shifting a driver's attention toward the in-vehicle systems [66, p. 651].

1.2.2 Vehicle Operation

The continual development of electronic components (miniaturization of integrated circuits, rising computation power and thus falling microprocessor prices [69], [70], etc.) has led to products such as function-rich navigation systems, complex infotainment or entertainment devices, Advanced Driver Assistance Systems (e. g. parking aid, electric power steering systems, collision warning or cruise control systems) [64, p. 217], [71], [72], which have turned the vehicle into a technology-centered multifunctional environment [25]. These new functionalities have been added to vehicles' state-of-the-art *comfort controls*[10], leading to additional driver diverson.

For example, the use of ADAS aims to "partly support and/or take over tasks of the driver" which means, in other words, to generally provide safer driving conditions [64, p. 216], [63]. Furthermore, these systems are supposed to be able to detect a critical driving situation and to inform the driver so that a "compensatory maneuver" can be performed [35]. It is incontestable that these information and communication systems could be the source of decreased driver performance [74] and subsequently increasing crash risk [65], [66] caused by information overload, driver confusion and distraction from the primary driving task (by violating the two most important automotive paradigms "*driver's eyes have to be kept on the road*" and "*driver's hands have to be kept on the steering wheel*" [67, p. 28]).

Intelligent and user-friendly designed interfaces create possible remedies for the indicated problems by assisting the driver not only with the primary driving task[11], but also supporting him/her (in a convenient and non-distracting way) in the secondary/tertiary tasks[12] [65, p. 256], [48, p. 25].

1.2.3 Conversation and Telecommunications

Aside from the factors of distraction directly resulting from vehicle handling, a number of frequent side activities such as conversation with passengers, cell phone or PDA operations, eating/drinking, personal hygiene (fixing hair, applying make-up, shaving, brushing teeth, biting

[10] "Comfort controls" are devices which are not required for handling the car – such as window and sunroof controls, door locks, interior lighting, mirror control, etc. [73].

[11] Driving is here considered as a cognitive task composed of the four main subprocesses (*i*) perception, (*ii*) analysis/evaluation, (*iii*) decision, and (*iv*) action . To be performed, each phase presumes the achievement of the previous one [35].

[12] In the automotive domain, the driver's activities are divided into primary (driving related), secondary (car status, navigation), and tertiary tasks (comfort and entertainment services) [75], [76].

nails, etc.), and others have been identified as potential causes of accidents [77, p. 436f], [78, p. 8].

Conversations

According to a study presented by Stutts *et al.* [79, p. 42], 77.1% of all drivers are engaged in the activity of conversations with passengers while driving, but at the same time numerous studies have proved that most car drivers are using their vehicle alone:

(*i*) In Scotland, 85% of the people are using their car alone when commuting between home and work; another 11% of cars are occupied by two persons (this corresponds to a average number of 1.20 persons per car)[80, p. 20f].

(*ii*) A Swedish study reports that the number of persons per car is on average 1.60 on long journeys (above 5 kilometers), compared with 1.30 for short journeys [81, p. 9].

(*iii*) Car occupancy surveys were undertaken throughout the greater Manchester area in the years 1989 to 2005. In this span of time, the average occupancy declined from 1.31 to 1.19 in peak hours (8.00 – 9.00 AM), and from 1.40 (in 1994) to 1.30 in off-peak periods (10.00 – 12.00 AM). These values correspond to a change in single car occupancy from 76% to 84%, and from 69% to 75% respectively [82, p. 59f].

With consideration of these results the class of conversation activities can be disregarded (because of little contribution overall).

Cell Phone Usage

It has been stated, for instance in [79, p. 42], that more than one third (34.3%) of car drivers are using their cell phone while driving. Dingus *et al.* [77, p. 129f] indicated that the highest level of inattention (from secondary tasks) causing crashes occurrs from drivers talking on their phone (8%), followed by drivers talking to passengers – the high contribution of the cell phone as a distraction item justifies a closer scrutiny of its usage.

National studies in the United States have shown that 5.4% of the drivers observed during daylight hours are holding a cell phone, and another 0.6% are using a hands-free device. Among drivers who own cell phones, 73% say they talk on the phone while driving for at least some of the time [83, p. 3]. According to a 2006 study by the National Highway Traffic Safety Administration (Glassbrenner *et al.*, [84]), the overall hand-held cell phone usage in the U.S. decreased to 5% in 2006, from 6% in 2005 (and to 8% from 10% for drivers in the age-range from 16–24 years)[13]. The hand-held phone use rate translates into 745,000 vehicles on the

[13]The decrease results from a "hand-held cell phone use while driving" ban by law.

roads in the U.S. which are driven – at any given daylight moment – by someone talking on a hand-held phone.

As many of the car drivers use their cell phone while steering, and due to the fact that talking on cell phones has an essential impact on road safety, the danger potential of cell phone distraction must be investigated and suitable solutions to compensate for them must be established.

Distraction Types

Two main sorts of risks of the coexistence of vehicle driving and cell phone usage have been identified: (i) the driver must take his/her eyes off the road while dialing and (ii) the driving person is fully engaged by the cell phone conversation so that his/her ability to concentrate on the task of driving is heavily impaired [85].

Landsdown *et al.* [86] investigated the general impact of side activities on the primary driving task and found that they led to significant compensatory speed reductions (performing multiple secondary tasks influences vehicle performance adversely – demonstrated with significantly reduced distances travelled and increased brake pressures).

Over the last years, the penetration of wireless communication devices and hands-free phones has risen continuously. Bluetooth, which is the basic technology used for virtually all any wireless communication in vehicles, had a penetration level of close to 20 percent in Europe in 2005 and is expected to reach 90 percent in 2010 [87]. Dünnwald [88] cites a study from In-Stat/MDR (11/2004) where the Bluetooth penetration level is estimated to reach 64.2% by 2008. A survey commissioned by Transport Canada found that 78% of all cell phone users in a car use hand-held phones and moreover, that 81% of the interviewed persons believed that cell phone usage poses a safety risk [89] (another survey performed in Canada has confirmed these results [90]). Distraction from both hand-held and hands-free phones will initially be considered separately, although there is evidence that the level and nature of distractions from their usage are rather similar.

(i) **Hand-held cell phones:** It has been proven that the use of hand-held cell phones generates multi-level distraction, e. g. by requiring (i) the driver's eyes to locate a ringing phone, (ii) the driver's hands to hold the phone or dial, (iii) the driver's ears to listen to the call, and (iv) the driver's attention to carry out the actual conversation [91, pp. 4–6]. The National Highway Traffic Safety Administration (NHTSA) [92] estimates that about 10% of drivers on the road at any given daylight moment are using their cell phone. Another study from Sundeen *et al.* [93] estimates that 73% of the USA's cell phone subscribers use their phone while driving. In [94] and [95] Alm *et al.* and the NHTSA have shown that the reaction time of drivers using cell phones while driving increases by 0.5 to 1.5 seconds – with the consequence that drivers have difficulties maintaining the correct position or the appropriate speed on

the road. Redelmeier and Tibshirani [96] have found evidence that drivers who use hand-held phones face a risk of crash which is four times higher than for drivers not using a cell phone. In response to these concerns, several countries all over the world and some US states prohibit hand-held cell phone use while driving [91]. The effect after banning hand-held use of cell phones in the United States on overall fatalities is ranged between reductions of 9% and 21%, depending on how long the law has been in effect [83, pp. 11–13]. The reported rates of traffic accident reduction will perhaps be outperformed by the state of Ontario, Canada, where the government has included iPods, portable DVD players and laptop computers in the ban [97].

(ii) **Hands-free cell phones:** Hands-free devices eliminate some issues regarding operating the car [98]. But on the other hand, studies have shown that cognitive demands of conversation are not eliminated, and therefore distraction still occurs [99], [97]. Morrison-Allsopp has presented a study which shows that using hand-held devices causes a four times higher number of accidents than using hands-free devices [98, p. 9]. A review from the NHTSA [95] found consistent evidence in that manually dialing using a hand-held cell phone reduces drivers' traffic awareness. Furthermore, the report shows that using hands-free phones has less effect on improving drivers' vehicle control performance, but increases their reaction time and concurrently decreases their awareness. McCartt and Hellinga [100] have shown that there is only a minimal difference in distraction when using hand-held or hands-free phones. A recent study by the Ontario Medical Association [101] claims that there is no difference between hands-free and hand-held cognitive distraction.

These findings indicate that (*i*) governments should not only ban hand-held phone calls but also restrain drivers from hands-free phoning, (*ii*) drivers should be warned about mistakenly thinking they are safe, (*iii*) assistance systems would be useful for detecting phone calls and reacting to them (e. g. by taking the vehicle control, stopping the car, delivering warnings), etc.

1.3 Distraction Forecast

One of the reasons for the evolution of existing and the deployment of new vehicular interfaces and applications is their potential to reduce the driver's cognitive load[14] [103]. A steadily rising number of devices in the near future is confirmed exemplarily by the following statistics:

[14]This statement is confirmed, for instance, in the IST Call 7, Action Line 1.5.1 – Intelligent transport systems, "Focus is the development of ADAS and in-vehicle multimedia platforms which take into account the driver's cognitive load in an optimal way [..]" [102].

(*i*) Strategy Analysis International[15] predicted that the global car electronics market will increase from $40 billion for 2007 to $60.8 billion in 2013, an increase of 52%[16].

(*ii*) It is estimated that costs for software and electronics in a vehicle will exceed the 50%-margin of overall car-manufacturing costs by 2015 [31].

(*iii*) There are estimates that already today more than 90% of all vehicle innovations are centered on IT (hardware and software) [31].

To avoid distraction due to information overload or technical complexity, devices, applications and gadgets have to be analyzed regarding means of operation and – if necessary – adapted to the needs of drivers, so that they are not actively required to cooperate with the interface or participate in interaction, and can thus keeg their attention on the main task of driving. It is a major goal to provide means for reducing drivers' level of cognitive load, particularly in situations where secondary tasks, such as regulating the air-conditioning system or tuning the car stereo, are performed, while the necessary rate of feedback on primary tasks is still delivered to the driving person. The challenge is to find a well-balanced trade-off among driving comfort, ease of interaction, and safe vehicle operation.

[15]http://www.strategyanalysis.com, last retrieved August 30, 2009.

[16]EETimesAsia (eMedia Asia Ltd.): New markets, technologies energize car electronics, http://www.eetasia.com/MAGAZINE/EDNOTE0809B.HTM, last retrieved August 30, 2009.

2 Perception

Humans recognize their environment by means of ten or more senses (vision, hearing, touch, smell, taste, balance, temperature, pain, proprioceptive sensibility, "sixth sense", etc.). A rough outline – taken and adapted from Silbernagel [17] and Bernadine Dias' lecture on "Robotics and Artificial Intelligence" [104] – is given in Table 2.1 below.

No.	Human sense	What is sensed?	Used sensor	Reference
		Traditional Senses		
1	Vision (seeing)	Electromagnetic waves	Eyes	[105], [106], [107]
2	Audition (hearing)	Pressure waves	Ears	[105], [106], [108]
3	Tactition (touch)	Contact pressure	Skin	[105], [109]
4	Gustation (tasting)	Chemicals (Flavor)	Tongue	[105], [110]
5	Olfaction (smelling)	Chemicals (Odor)	Nose	[111], [112]
		Physiological Senses		
6	Thermoception	Temperature (Heat)	Skin	[113]
7	Nociception	Physiological pain	Skin, Organs, Joints	[114], [115], [113]
8	Equilibrioception	Sense of balance	Ears	[116]
9	Proprioception, Kinesthesia	Body awareness	Muscles, Joints	[117]
10	Time	Cognitive processes	Combination of Senses	[118]
11	Extra-sensory perception	"Sixth Sense", ESP	None	[85]
12	Magnetoception	Body direction	Non-human sense?	[119]

Table 2.1: Humans' traditional and physiological senses.

Human senses are typically classified into two groups, *Traditional Senses* and *Physiological Senses* [105, p. 278], with the focus being on traditional senses for this research work.

When analyzing the traditional senses – which can be perceived consciously or subconsciously – with respect to their applicability in vehicles or to driving situations in general, it can be determined that these senses are of different importance. Bernsen [42, p. 2] indicated for instance, that "in the foreseeable future, intelligent multimedia presentation systems will mostly be using three modalities, i. e. graphics, acoustics and haptics". The utilization of other sensory channels in the automotive domain is ill-suited – *gustation* for example is not applicable at all (there is no associated connection between *taste* and activities involved in steering a car) [25, p. 78]. The same applies to the sense of *smell*, although there are some properties in the vehicular context perceivable by the *olfactory* sense (e. g. environmental scents, smell of burning or of leaking oil/petrol). Another issue related to the modalities *smell* and *taste* is the ever existing problem of the composition of odors and gustatory substances (a more exhaustive discussion on the potential and problems of smell and taste is given in the Paragraph "Potential of Smell and Taste" on pp. 43).

Consequently, the three senses (*i*) vision, (*ii*) hearing, and (*iii*) touch remain left for using in in-car applications to signal or notify the driver.

2.1 Vision

Vision is the main input channel for maneuvering a vehicle [75, p. 130], and – according to Mauter and Katzki [25] – up to 90% of all sensory input is received via the eyes. Dahm [50, p. 41] stated that about 80% of the information is delivered over the visual channel, and another 15% via the auditory channel. Recognizing persons, vehicles or obstacles in front of the car, or constantly burning or flashing lights in the dashboard are typical perceptions for the modality *vision*; however, slight changes in illumination (such as glaring sunlight or changing light conditions when driving through a tunnel) produce large changes in object appearance [54], which result in missing visual notifications (e. g. small bulbs lighting up on the dashboard) or lead to unintentionally ignoring traffic situations (a changing traffic light, or a car braking ahead).

2.2 Hearing

Environmental noise or motor sound, voice instructions from a navigation system, "beeps" from a Park Distance Control (PDC) system, etc. are typical stimuli for this channel of sense. Auditory notifications are often missed due to other, louder sources of noise such as the car stereo or "heated debates" with a passenger. On the other hand, sound output quickly becomes annoying since it disturbs other auditory activities like music or conversations [75, p. 131]. Warning signals normally have to be understood immediately, but in the case of spoken warning signals the driver has to listen to the end of the message in order to avoid misinterpretation. This characteristic or drawback of speech interfaces can be compensated to some degree by choosing short and concise messages. Furthermore, the message length is an important factor, for instance when incorporating multimodal interfaces (e. g. a combination of visual and auditory stimulation) it has to be ensured that both types of notification can be perceived in nearly the same span of time (see Fig. 2.1 for an illustration). It should be noted that it would be better to use also binary signals (instead of spoken ones) for auditory notifications, e. g. a "beep" on the left audio channel for informing about a "turn left" request.

A more detailed discussion of these issues is given on p. 77, in Appendix D.1: "Human Reaction Times" on p. 225, or in the experimental section for the study on "Simulating Real-Driving Performance" on p. 161.

Sound can also be perceived by means of vibrations, conducted through the body by the sense of touch (in particular, frequencies below or above the audible range can be detected as vibrations only).

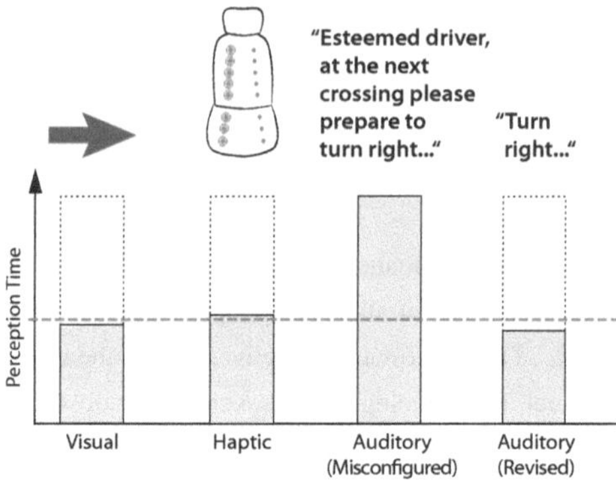

Fig. 2.1: Perception times of individual feedback channels have to be aligned to one another in order to get meaningful results.

The potential of Vehicle-2-Driver and Driver-2-Vehicle communication with speech information (as, for instance, voice commands for both input and output in navigational devices) is of increasing importance for the support of the more complex systems in a car [120]. However, it has to be considered that auditory output interferes with the driving capabilities to a certain degree [75, p. 131], and that speech-based input is error prone and therefore not the best modality for all control activities [121, p. 229]. In [122, p. 219], Basapur *et al.* reported that users spend only 25–33% of the time on text input, and the rest of the time on error detection and correction – while trained Automatic Speech Recognition (ASR) systems can achieve about 3% of Word Error Rate (WER), WER can increase to 40–45% as acoustic conditions degrade (engine or environmental noise, communication of passengers, etc.). Past experience has also shown that drivers do not like to receive driving instructions from a voice command system [31] and that speech modalities in native or known languages have very high saliency – the last property might distract a driver's attention from the traffic in situations where attention to the traffic has top priority [123].

2.3 Touch

This is the sense of *force perception* on the skin and haptic devices can be seen as a special form of a force display [124]. Beside vision and hearing, the sense of touch is considered the most important sensory channel for future vehicular applications and has the potential to improve driving safety, enhance driving experience, and others [125].

Up to now, most Human-Computer Interaction systems focus primarily on presenting visual or, to a lesser extent, auditory information [126], [127] – this is also the case in current vehicular applications. When developing control elements, people often forget that haptic interfaces have the potential to increase the quality (or performance) of Human-Computer (or Driver-Vehicle) Interaction by accommodating the sense of touch [128]. But, as investigated elaborately for instance in [129, p. 126], visual information can influence haptic perception and moreover, it is believed that visually presented information overrides touch information when they are presented simultaneously [130, p. 24]. Since the visual display is an integral part of any vehicular interface, it is important that the presentation of visual information is congruent with the haptic stimuli and scheduled at appropriate times. Furthermore, it has to be considered to what extent vision can influence and/or impair vibro-tactile perception [129].

Lin *et al.* [131] have analzyed *haptic* interfaces in traditional applications and found that they provide an attractive augmentation to visual displays and enhance the level of understanding of complex data sets. Transferring their findings to the vehicular context opens up the possibility of supplementing the other channels of sense (normally *vision* and *hearing*) with *haptic* interfaces, thereby reducing the required level of user attention (LOA). This assumption is supported by findings from a recent study by van Erp *et al.* [132], which provides evidence that drivers experience a reduced (subjective) workload when using tactile instead of visual navigation displays. Furthermore, results from Ho, Spence, and Tan [133] reinforce that vibro-tactile signals are eligible to convey information reliably from the vehicle to the driver.

Summary and Impact

For delivering notifications, warnings or general information from the vehicle to a driver, the utilization of the three sensory modalities, *vision*, *hearing*, and *touch*, is appropriate. All of them have weak points and all of them depend to a greater or lesser degree on drivers' characteristics like age or gender.

There is evidence that the majority of information (in vehicles) is transferred via the *visual* sense (80 – 90%), followed by the *auditory* sense (approximately 15%). The sense of *touch* is still underused, but has great potential for supporting the other two modalities (e. g. when applying multimodal interfaces).

In anticipation of the results of this work and with respect to findings from other research studies, it has been shown that the response time from *vibro-tactile* stimulations is similar to the reaction time from *visual* or *auditory* notifications (and often performs even better). Furthermore, it has been proven that *vibro-tactile* stimuli delivered via the car seat are perceived implicitly, and thus would cause no additional distraction of the driving person.

3 Driver Expression

The second interesting field of investigation in vehicular interfaces is the articulation of commands or messages from the driver towards the car and its assistance systems. A driver is an active constituent in the interaction loop and is continuously requested to react to the perceived information from the environment or from the vehicle's systems.

As stated in "Vehicle Operation" on p. 9, driver activities are grouped into primary, secondary, and tertiary tasks. For the investigation of driver-vehicle expressions, this classification of activities needs to be revised to differentiate only between two classes: (*i*) primary or vehicle- and steering-related activities such as steering, operating multimedia device controls, route planning, navigation, etc. and (*ii*) secondary (or driving independent) tasks such as person-related services or service classes (mobile communication, Internet-based applications, intra-car communication, etc.).

The most prominent recent research challenges for the latter category are:

(i) **Personalized services:** (*i*) Authorization (permit or deny engine start, configure specific motor settings, horsepower regulations, personalized settings for seat, mirrors, or radio stations, personal car insurance policy, cause-based CO_2 taxation), (*ii*) security services (secure transactions for mailbox access, mobile banking, and online-shopping while concurrently preserving codes or passwords from being spied on by passengers), and (*iii*) safe-driving or "Fitness to Drive" services (monitoring a driver's behavior and comparing it to reference data; permitting or denying actions if the driver is in an inappropriate condition, e. g. fatigued or drunk).

(ii) **Communication:** (*i*) Intelligent networked car solutions (cause-based taxation, personalized insurance policies, logistics and fleet management), (*ii*) next generation Driver Assistance Systems (DAS), (*iii*) vehicle comfort applications (auto-configuration at boarding time for any vehicle, particularly for shared cars), (*iv*) real-time traveller and traffic information, etc. For further information on this class see e. g. the "Intelligent Car Initiative" ([134], [135]).

Traditional control activities are most frequently executed explicitly, while the person-related or Internet-enabled services are typically initiated implicitly, meaning that they require no active task cooperation from the driver. In order to implement personalization in vehicles, reliable, trustworthy, unambiguous and continuous identification mechanisms have been discovered as a key enabling technology (Heide *et al.* [136]).

The steadily increasing number of DAS and other appliances (e. g. infotainment systems, or communication devices) in vehicles, together with a rising operation complexity of such systems, have further accelerated the demand for distraction-free and/or implicit Driver-2-Vehicle articulation capabilities.

3.1 Modes of Interaction: Explicit versus Implicit Feedback

In today's Driver-Vehicle Interfaces, user feedback is most often given *explicitly*, requiring the active cooperation of the driver (e. g. by pressing a button, activating a switch, articulating a command for the voice recognition system, tilting and turning the head to be recognized by the face recognition unit, etc.) which potentially causes an increase in his/her cognitive load. For future Driver-Vehicle Communication systems, a new way of driver articulation is recommended, in which the system has the ability to detect, interpret, and react to a user's demand or requirement by utilizing special sensors.

The opportunity for providing *implicit* input in vehicles would be given, for instance, by pressure sensors unobtrusively integrated into the driver seat, and continuously monitoring a driver's sitting postures (this approach is introduced and evaluated in the Section "Experiments", starting on p. 119). According to the conducted tests and previous findings, posture pattern analysis seems to be a suitable method for driver identification and/or activity recognition.

3.1.1 Explicit Interaction

A major challenge in the design of a new generation of Driver-Vehicle Interfaces is negotiating the transition between *explicit* and *implicit* interaction [38]. Currently, most of the interfaces require an explicit style of interaction (which is often error-prone, inconvenient, or even unsuitable) – the driver is actively participating, issuing specific directives to the system, and expecting response from it ([37, p. 2], [38, p. 2], [137, p. 1]). Explicit interaction always needs an interchange between the user and a particular application or system the user is interacting with – Driver-Vehicle Interaction is centered on the system aspect, while the user focus is on the communication interface – awaiting an response from the vehicular system.

Schmidt [138, p. 184] defines the basic procedure for user-initiated explicit interaction as a three-stage process, based on the "execution-evaluation cycle" introduced by Norman, 1988: (*i*) the user requests the system to perform a certain action, (*ii*) the action is carried out by the system, either with or without feedback on the process, and (*iii*) the system responds with a reply if necessary or appropriate.

The process of explicit interaction is thereby similar across the utilized interaction modality (or modalities); only the "views" of presentation and interaction vary according to the selected modality.

Contemporary *explicit* Driver-Vehicle Interaction systems primarily operate on *speech* ([139], [52]) or *vision* recognition methods ([140], [141]). Each sense has its benefits and drawbacks

[53, p. 374]; for instance, the modality *speech* has been considered the most convenient in-car interaction modality [139, p. 155]. However, later research revealed that speech recognition failed to live up of its potential, particularly in the automotive domain, because voice is influenced by a multitude of parameters including user characteristics like age, gender, mental workload or emotional stress and in addition, voice interaction may distract the user from the task of driving [139, p. 156]. Finally, ambient noise is responsible for distortion of spoken instructions [53, p. 368].

3.1.2 Implicit Interaction

In ordinary human-to-human communication, a lot of information is exchanged *implicitly* by using body language, gestures, inflection of voice, facial expressions, etc. The combined usage of such *contextual* information, together with traditional feedback, increases the robustness of human-to-human communication by redundancy (for example, a concerned or disagreeing nodding while answering "yes" or "no") [37].

However, prior to considering the means for more natural and fluent interaction capabilities in vehicles, and ways to enable or enhance interaction with computer systems in vehicles (with respect to the fact that usual desktop interaction paradigms are often inappropriate or unavailable [44]), related terms have to be defined. To allow the driver to focus on the primary task (driving), the Driver-Vehicle Interface is required to be in the background; this makes *implicit* interaction the preferred technique, as for instance proposed by Rani *et al.* [127] or Witt and Kenn [142] as "[..] implicit interaction is unobtrusive, non-distracting and lets the person stay in his/her current state or activity".

Implicit Human-Computer Interaction (IHCI) in the context of this work is defined following the rough definition from Wilson and Oliver [44, p. 1] as "[..] interactions that neither require the discrete inputs of explicit interactions, nor give the user an expectation for a discrete timely response", and enriched with the considerations of Schmidt [138, pp. 189]: IHCI is the "[..] acquisition of *implicit input* from the user, maybe followed by the presentation of *implicit output* to him/her", where *implicit input* allows a system to perceive users' interaction with the environment as well as the global scope where an action takes place, and *implicit output* is not directly related to an explicit input, and is seamlessly integrated into the environment and users' tasks. Furthermore, implicit HCI is "[..] an action, performed by the user that is not primarily aimed to interact with a computerized system, but which such a system understands as input" [37], i. e. the computer understands what the user is doing and knows the consequences [142].

It should be noted that interaction in *implicit* Driver-Vehicle Communication systems is vulnerable to a multitude of problems, such as unintended actions, undesirable results or the impossibility of correcting input errors [38].

These problems are mostly caused by (*i*) a mental absence of the user, (*ii*) the concentration on other tasks, or (*iii*) a driver's intention misunderstood by the system. It is absolutely essential to consider and resolve these issues when designing in-car interaction interfaces that operate implicitly.

3.2 Modes of Interaction: Cardinality of Sensory Channels

Apart from the interaction types *implicit* and *explicit* it is important to classify and distinguish interaction systems based on the cardinality of employed sensory modalities, which are either *unimodal*, the general case *multimodal* (bimodal, trimodal, etc.), or the special case *amodal*.

3.2.1 Unimodal

For Vehicle-2-Driver (V2D) feedback today, the two sensory modalities *vision* and *hearing* are mostly used in a standalone (*unimodal*) manner, meaning that information is either presented as visual stimulus, e. g. as a pulsing lamp, or as auditory signal, for instance as a "beep" of the Park Distance Control (PDC) system.

The recent emergence of information on and around the dashboard, as well as the potential misuse of a sensory modality for a specific interaction demand, has led to a high cognitive load of the driving person and consequently to information oversaturation and distraction. It is indicated that these factors of distraction on the one hand, and road traffic accidents on the other, are in a direct relation to each other. Exemplarily, the results shown by Paulo *et al.* [143] stated that distraction is a significant determinant of traffic accidents.

In 2001, the European Commission proposed the goal of saving 25,000 lives annually on European roads by the target date of 2010 [144] (today, 1,300,000 accidents a year cause 40,000 deaths and 1,700,000 injuries on the European roads) – the European Parliament and all Member States signed this contract. This attempt demands solutions, such as, for instance, (*i*) the introduction and disposition of new sensory modalities (*touch*, *smell*, or *taste*), (*ii*) a notification via two (or more) modalities simultaneously (this option is supported by the findings of Vilimek *et al.* who stated that a single modality does not allow for efficient interaction across all tasks, while multimodal interfaces enable the user to (implicitly) select the best suited [145]), or (*iii*) a more autonomous operation of Driver Assistance Systems, perhaps combined with a reduced amount of feedback.

3.2.2 Multimodal

Humans perceive their environment not over a single information channel but as a combination of several input and output modalities [145] (=*multimodal)*, including *vision*, *hearing*, *touch*, *gustation*, *olfaction* (and possibly others) [146].

Multimodality has the potential to (*i*) facilitate richer interaction styles in Human-Computer Communication [57], (*ii*) release other involved sensory channels from high or excessive cognitive workload [58, p. 843], (*iii*) increase the usability by compensating for the weakness of one modality by the strengths of another (for example, on a small visual display in the dashboard, it would be quite difficult to read a larger quantity of information, however very easy to listen to it) [146], and (*iv*) increase the information processing capacity of the user (as defined in Wicken's Multiple Resource Theory (MRT), [59]).

Prewett *et al.* [147] investigated the benefits of feedback delivered through multiple modalities instead of a single modality. The overall results indicate that visual-tactile bimodal feedback provides significant advantages over the simple unimodal visual feedback and suggests the utilization of the tactile modality in addition to the visual modality when providing task information. According to Oviatt [56, p. 76], users have a strong preference for multimodal interaction, however, there is no guarantee that they will *always* communicate in this mode. A user study addressed in the same work has shown that different users expressed their commands *multimodally* only 20% of the time, with the rest using just one sensory modality. Pascoe *et al.* [148] introduced the term Minimal Attention User Interface (MAUI) as an interface mechanism that minimizes the amount of user attention that is required to perform a particular task and found that the complementary use of interaction modes is the key to successful MAUI design[17].

The challenge here is to discover new means of employment for each interaction modality in such a way that the original disadvantages of interaction techniques do not appear any longer.

Information among people can be conveyed in a number of ways. But for any human-machine communication it is insufficient to simply replace humans as information providers by machines (machines do not have the ability to express themselves with physical activities like gestures or facial expressions; a successful design of multimodal systems will therefore require guidance from cognitive science theories [56, p. 75]). The determined shortcomings of machines are reflected by multimodal interaction research efforts up to now: The bigger part has focused on using each modality separately or in pairs; minor research has been carried out on the basis of a detailed analysis of (human-human) multimodal interaction [57, p. 2].

As drivers must perceive many different types of information, a system exclusively using one modality might lead to cognitive overload – in virtually every circumstance carefully designed *multimodal* displays appear to be more beneficial than any *unimodal* display [126]. Another aspect legitimizing *multimodal* interfaces is situations or constraints where a dedicated channel of information is not available, or not suited very well. To give an example, drivers of cars are generally required to be fully aware of the environment, but there are a number of restrictions when using only one out of the two common information channels *vision* and *hearing*, e. g.

[17] In more detail they stated that minimal attention in user interfaces is achieved through the use of modes of interaction that distract or interfere least with the mode the user is already employing [148, p. 432].

dazzling or glaring sunlight, or limited visibility in foggy situations for the *visual*, or noise stemming from the motor or the environmental traffic for the *auditory* sensory channel.

In general, information should be presented as simply as possible. In the case of complex information, a *multimodal* display would be able to compensate for the workload of a driver to a certain degree, resulting in a shorter response time or an enhanced driving performance [126]. Siewiorek *et al.* [149] developed a multimodal car-interface integrating vision, speech, gestures, and touch that enables drivers to be more productive while driving, without making driving unsafe. Vilimek *et al.* [150] and Pieraccini [151] determined key factors for a successful design of advanced multimodal in-vehicle interfaces, and Pieraccini *et al.* [152] evaluated a prototype of a multimodal interface in a concept car and found that a deliberate user interface design allows expert users to speed up their interaction.

The introduction of a multimodal interface comprising the sense of *touch* would provide additional information through redundancy (a hazard would not contingently be seen in the presence of dazzling sunlight or dense fog, but would alternatively be felt with a specific vibration pattern applied on the seat). All the information between a driver and vehicular assistance systems is exchanged via the closed feedback loop shown in Fig. 1.1 (p. 5). As this feedback loop is susceptible to a number of disturbances it is feasible that the driver receives incongruent notifications from the different utilized sensory modalities. However, humans can perceive even incongruent bimodal (e. g. *visual, haptic*) information without any conflict [153, p. 314].

But redundancy is not the only benefit of multimodal interfaces; it furthermore would allow information to be optimally provided or distributed by using the best suited modality (based on the relation of the information which has to be transmitted and the available sensory channels). This approach requires an evaluation of the advantages and limitations of each sensory modality [1, p. 19].

Fig. 3.1 (adapted from [1, p. 44]) gives an overview of the appropriateness of combining modalities as redundant information sources (it is noted that the table is not symmetrical along the main diagonal). The highlighted regions (1 to 4) in the table are of special interest for this research work and can be interpreted as follows:

(1) Visual and auditory information channels complement each other in a nearly optimal manner. Both of them can be used as the primary or alternative source of information, resulting in the same quality of information processing.

(2) The haptic modality (as a combination of tactile and kinesthetic senses) is most qualified as an additional output channel, with the visual modality (still or moving images) as a primary information channel.

(3) and (4) Olfactory and gustatory are ill-suited as primary output channels, and remain poor, even when used as secondary information channels (this holds true for all the primary modalities *vision*, *hearing*, or *touch*).

Primary Modality	Alternative Modality: TE	GR	AN	SP	NS	TA	KI	TM	OL	GU
TEXT (TE)		-	n	++	n	n	-	--	--	--
GRAPHICS (GR)	++		++	+	n	++	++	--	--	--
ANIMATIONS (AN)	++	+		+	+	+	+	--	--	--
SPEECH (SP)	++	+	+		n	-	-	--	--	--
NON-SPEECH (NS)	+	+	++	+		-	-	--	--	--
TACTILE (TA)	n	-	-	n	+		+	--	--	--
KINAESTHETIC (KI)	n	--	-	n	-	+		--	--	--
TEMPERATURE (TM)	++	+	+	++	--	-	-		--	--
OLFACTION (OL)	n	-	n	n	--	--	--	--		++
GUSTATION (GU)	+	-	n	+	--	--	--	--	++	

Symbol Description:
++ very well suited as alternative output channel
\+ good enough, although not optimal
n neutral
\- not very well suited, but possible
-- hardly possible or even impossible

Fig. 3.1: Appropriateness of combining modalities as redundant information sources (adapted from European Telecommunications Standards Institute (ETSI) [1]).

3.2.3 Amodal

Multimodal interfaces incorporate different senses to provide adequate input or output for human-device interaction. But this approach is problematic, e. g. for people with sensing disabilities or for other limitations in interface operation. For these cases, Yfantidis *et al.* [154] proposed to use *amodal* information, where the commonly used sensory modalities are only employed as peripheral aids (they evaluated a prototypical system in an experiment with input of blind people on a touch screen and found promising results [155]).

Moreover, the term *amodal* denotes the absence and/or deviation of the perception conditions specific to a designated sensory channel (*vision*, *touch*, *hearing*, *smell*, or *taste*). Amodal information is information that is not specific to one sense, but can be detected across two or more sensory modalities – thus, amodal symbols are arbitrary or unconstrained, meaning that there is no natural or direct connection between a symbol and its representation [156, p. 578]. Amodal perception is also described as sensory imagination, for instance the back side of any non-transparent object is perceived only *amodally*. To fully understand the term *perception* itself, it is necessary to understand *amodal perception* (Nanay, [157]), where amodal perception is not limited to visual phenomena, but is very important in tactile sensory modality, too (humans for instance are *amodally* aware of those objects that they do not have any tactile contact with).

In the definition of Markovich [158], the phenomenon of amodal completion can be considered as a possible link between sensory and cognitive processes, because it refers to something that is at the same time (*i*) perceptual (we see that one vehicle partly overlaps another), and (*ii*) imaginative (the look of the hidden part is only suggested, but not explicitly given).

In the definition of Schwarz *et al.* [159, p. 64], the term *amodal* means to be fully abstract and completely detached or released from sensoric and modality-specific information. Changes in the intensity or the temporal and spatial aspects of sensory stimulation, including properties such as rhythm, tempo, duration, synchronicity, and collocation are examples of information not specific to a single sensory modality [160, p. 254]. Therefore, they are qualified to be presented as *amodal* information (rather than extracting a subset of a perceptual state and storing it for later use as a symbol, the amodal system transduces a subset of a perceptual state into a completely new representation language that is inherently non-perceptual [156, p. 578]). The interpretation of *amodal* information or symbols requires theories of combinatorics, reasoning, and/or inductive inference [161, p. 241], which potentially lead to different results for individual persons.

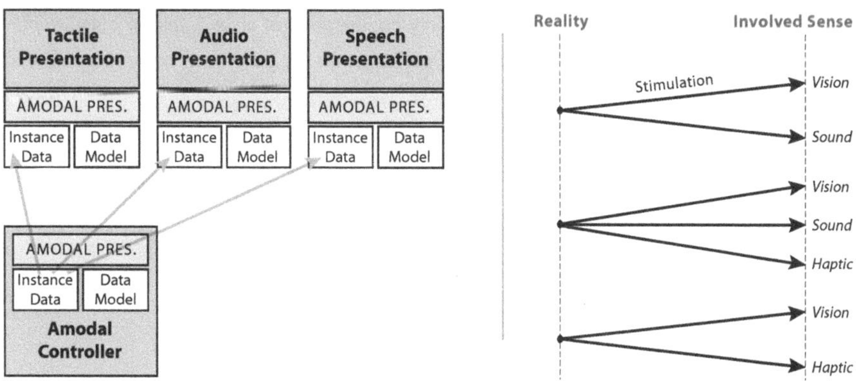

Fig. 3.2: The amodal specification assumes that a given aspect of reality redundantly stimulates different channels of sense (adapted from Stoffregen *et al.* [2]).

When considering the application of *amodal* information processing or presentation in the vehicular domain, the proposed solution has to cope with a number of challenges: The system should at least adapt to (*i*) a user's needs, (*ii*) the task scope, (*iii*) the utilized technical platform, and (*iv*) the context.

An agent-based interaction platform called "EMBASSI" that deals with the addressed challenges has been presented by Richter and Hellenschmidt [162]. Fig. 3.2 shows the part of the architecture responsible for amodal information presentation.

Summary and Impact

Future Driver-2-Vehicle (D2V) input will involve topics ranging from simple control commands over automatic driver identification tasks to the detection of a driver's ongoing activities. The research challenge in this area of application is the development of Driver-2-Vehicle interfaces that do not interrupt the driver from his/her primary driving task, e. g. allowing the driver to operate both hands-free and eyes-free[18], while continuously submitting data (for example to confirm his/her identity), transmitting control commands, or initiating activities.

Today's *explicit* vision and sound-based Driver-Vehicle Interaction systems exhibit the disadvantage of requiring active participation of the involved driver. This possibly leads to additional cognitive load, and in succession to distraction or stress, and hence to an increase in road traffic hazards.

Traditional interaction concepts have to be replaced by more intuitive, *implicitly* operating and distraction-free alternatives (whenever feasible), in order to ensure precise, accurate, efficient, and also convenient handling of future cars and DAS.

Regarding the cardinality of applied sensory channels, Wheatley and Hurwitz [163] recommended the use of *multimodal* interfaces (any combination of *visual*, *auditory*, and *haptic* modalities) instead of a *unimodal* information exchange for the purpose of achieving safe and effective future Driver-Vehicle Interaction.

[18]Hands-free and eyes-free operation in this context means that the hands remain on the steering wheel all the time and the eyes are exclusively focused on the road [139, p. 155].

4 Perception and Articulation

The state of the art in Human-Computer Interaction (HCI) is still the utilization of the *visual* and/or *auditory* information channels, often without reflecting on other possible ways of notification or expression. But from a general standpoint this choice is correct, as there is evidence that the biggest part of information (in vehicles) is delivered via these two senses (several studies confirmed that about 80% of all the information is perceived with the eyes, and another 15% with the ears, e. g. Dahm [50], Mauter *et al.* [25], Evans [34], or Verwey *et al.* [164]; Wierda *et al.* [51] have concisely underpinned these results with the statement "Driving is seeing [..]"). A more detailed investigation into the contribution rates of the individual senses is given in the Section "Motivation for Tactile Displays" on p. 46.

4.1 Application Domains

The vehicle is one of the most important devices humans interact with. Mostly as a result of technological advances, today there are countless in-car services offered to the driver. According to the terms of interaction with the driver, the environment, vehicular sensors, global information or service providers, etc. or based on drivers' demands like driving comfort, safety facilities, etc., the different services can be classified or mapped into distinct vehicular application domains, such as (*i*) control activities, (*ii*) navigation, (*iii*) safety services, (*iv*) maintenance applications, (*v*) communication services, (*vi*) entertainment and personal services, etc.

The classification of these application domains thereby benefits from the fact that automotive systems are "closed environments", specified, amongst others, by the following parameters: (*i*) the driver is required to sit on the driver's seat while steering the vehicle (and sitting posture and position are usually relatively fixed [165]), (*ii*) the interaction environment in a specific vehicle normally remains the same (not only during a particular journey, but almost over the lifetime of that vehicle), and might be easily enhanced with the state or context of the driver, (*iii*) with almost all appliances, devices and controls in a vehicle connected to the internal buses (e. g. CAN, LIN, MOST, FlexRay), a driver's task should be easily detectable and processable (by passing on relevant sensor data between different ECUs), and (*iv*) environmental data should be accessible via sensors in the car for measuring distances to objects in the surroundings and road conditions (e. g. wet, icy, slippery), and wireless communication technology to poll for activities/incidents in the environment (e. g. road accidents, traffic density, road works ahead), see Hoedemaeker and Neerincx [165, pp. 1].

The breakdown[19] into application domains allows for an "easy" review of best suited interaction capabilities on a higher, more generalized, level, which should also be applicable to services and applications emerging later.

4.2 Participation Demand

A driver's articulation desire includes, amongst others, (*i*) a response to vehicle messages or notifications, such as in-vehicle status information or danger warnings, (*ii*) a reaction to environmental conditions, e. g. pedestrians on the street, traffic jam or road work ahead, approaching snowstorm or loud engine noise, (*iii*) a spontaneous desire[20], such as changing the radio station due to advertisements, shifting the gear down in order to plan overtaking or blowing the horn due to a vehicle blocking the exit, and (*iv*) a selection or configuration of personalized services, such as calculating a personalized car insurance rate or cause-based taxation, applying person-centric vehicle settings (mirror and seat position, radio station, temperature of air conditioning system) and/or motor management (horsepower regulation, fuel composition).

It should be noted that particularly the fourth bullet point would require the capability of person identification and/or authentication in the vehicle.

			VEHICLE OUTPUT				DRIVER INPUT
	Perception					Articulation	
Vehicle / Driver Domains	VISION	SOUND	HAPTIC	OLFACTORY	GUSTATORY	EXPLICIT	IMPLICIT
VEHICLE CONTROL							
NAVIGATION							
SAFETY SERVICES	WELL RESEARCHED AND COVERED BY LITERATURE		RESEARCH FOCUS	STILL UNCLEAR AND SIGNIFICANCE UNKNOWN		WELL RESEARCHED	RESEARCH FOCUS
MAINTENANCE							
PERSONAL SERVICES							
ENTERTAINMENT							

Fig. 4.1: Overview of vehicle-driver (output) and driver-vehicle (input) interaction modalities versus application domains with highlighted regions of research focus.

[19]For broader information on this topic consult e. g. the publications of Tango *et al.* [64], Kantowitz and Moyer [6], Hulse *et al.* [166], Heijden *et al.* [167], Erp *et al.* [168], or van Driel *et al.*[169].

[20]A spontaneous desire in the context meant here is a situation which is not foreseeable by the system.

4.3 Interaction Modalities versus Application Domains

Considering the various input and output prospects between a driver and a car on the one hand, and the different interaction demands for vehicular services (consolidated into application domains) on the other, a tabular record should be best suited with regard to providing a structured and thorough comparison of these two dimensions.

Furthermore, this form of representation would offer the opportunity of comparing a certain driver/vehicle application domain with any perception modality or input channel. Fig. 4.1 gives an outline of the suggested comparison table. A completion of the table with specific applications/services and a mapping to the experiments conducted in this research work (presented in the Section "Experiments") is given in the next Section "Hypotheses and Research Questions".

5 Hypotheses and Research Questions

5.1 Interaction Modalities

The focus of this research work is on automotive interfaces and in-vehicle interaction. Out of the five human senses *vision*, *audition*, *tactition*, *olfaction* and *gustation*, the last two (smell and taste) provide only marginal contributions to Driver-Vehicle Interaction. Of the remaining three, the modalities *vision* and *hearing* are highly stressed[21] in today's human-machine interfaces – the utilization of the sense of *touch* is relatively uncommon and originally emerged from communication systems for the visually or hearing impaired [4], such as "Braille" or "Tadoma" (see appendix E: "Alphabets Related to Touch"). Further potential can be seen in the class of *physiological* senses, for example in the *vestibular* sensory system (also known as *balance*).

5.1.1 Articulation

Explicit input, which is common in today's Driver-Vehicle Interfaces, requires active participation of the driving person for (*i*) initiating actions (e. g. for manually switching controls or operating pedals) and (*ii*) identification or authentication tasks, such as looking into a retina scanner or putting a finger onto a finger reader. Unfortunately, this effort may generate additional cognitive load, which subsequently distracts from the primary task of driving.

Gesture-based interaction emerges into different fields of application in which an individual has to communicate with a computer or machine [170] (e. g. multimedia gadgets such as the Nintendo Wii or Apple iPhone). Operating vehicle assistance systems via gestures from hands or body, the head, or the eyes could be imagined; however, they might violate the "hands on the steering wheel" or to a lesser extent the "eyes on the road" strategies [168, p. 3].

A solution for the problem of excessive cognitive load is potentially found in the disposition of the modality *touch* for driver-vehicle input. Posture patterns implicitly captured on a driver's seat and backrest have the chance to revolutionize Human-Computer Interaction, particularly in the automotive domain.

Enabled by technological advances, Driver Assistance Systems are already aware of the vehicle's state; pressure sensing technology and pattern analysis methods allow a driver's condition to be determined – the combination of both parameters would enable a vehicle to react automatically to changing environmental and personal conditions.

[21] This is a good choice because there is evidence that the biggest part of information in vehicles is delivered via these two senses, e. g. in [25], [49]. For instance, for the visual sense, Verwey *et al.* stated [164] that "[..] the most dominant source of danger in vehicles is not looking in the appropriate direction".

5.1.2 Perception

In the automotive domain, many situations in which the *visual* and *auditory* channels of perception are highly stimulated are known. Under these circumstances, messages or warnings from the vehicle can suffer from inattentiveness due to information overload. A viable approach to solving this problem could be the additional employment of the sense of *touch* for (*i*) raising a driver's attention and (*ii*) reducing the high burden on the *visual* and *auditory* sensory channels.

Another aspect legitimizing the application of perception based on the sense of *touch* is the compensation of limitations of the visual (e. g. reflecting sun on crossings controlled by traffic lights, changing light conditions when driving through a road tunnel, poor visibility in foggy situations, bad weather) or auditory senses (engine noise, superimposition of voices in communication with passengers or while using the cell phone, etc.).

The chosen approach of implicit vehicle-driver notifications opens new perspectives for the task of driving, such as a decline in road accidents and casualties, a discharge of cognitive load, and finally operator convenience and driving comfort.

5.2 Research Questions

The scope of this work can be defined by the two terms *Driver-Vehicle Input* and *Vehicle-Driver Output* as described above. The interrelationship between the research focus and the overall Driver-Vehicle Interaction demand is depicted in Fig. 5.2 on p. 33).

It should be noted that the vehicle-driver domains shown in that figure are not separated strictly – a crosslink among the different domains is conceivable, for instance between the classes "Navigation system" and "Entertainment". Furthermore, it is important to remark that the classifications and their associated actions may be tentative in a few places.

5.2.1 Objective

The goal of this research work is to give substantiated suggestions on how to use the sense of *touch* (if applicable at all) in the automotive domain for the purpose of reducing a driver's cognitive load (or compensating for his/her distraction).

Accordingly, several studies in the two domains (*i*) input and (*ii*) output have been conducted, considering issues like (*i*) the feasibility of sitting posture patterns for identification tasks, correlation between dynamic sitting postures and driving activities, etc., and (*ii*) interaction performance of the different sensory modalities (vision, hearing, touch), interrelationship between age and/or gender and reaction times, etc.

5.3 Experiments

In all the experiments and user tests conducted in the scope of this research work (for an overview of studies related to perception (output) and articulation/expression (input) see Fig. 5.1) only systems comprising a single driver and one car have been considered.

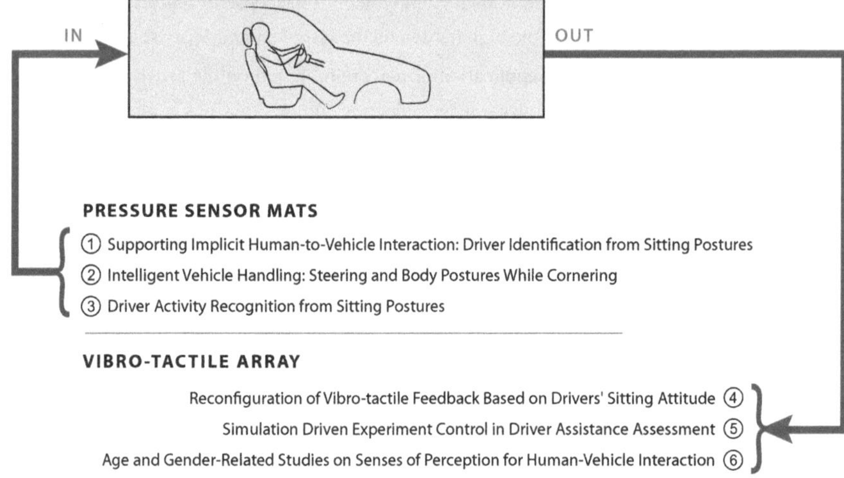

Fig. 5.1: Experiments conducted for investigating driver expression and perception.

	VEHICLE OUTPUT					DRIVER INPUT	
	Vehicular services (VS) involve the following senses of perception					VS involve means of driver articulation	
Vehicle / Driver Domains	VISION	SOUND	HAPTIC	OLFACTORY	GUSTATORY	EXPLICIT	IMPLICIT
VEHICLE CONTROL: Steering (accelerate, brake); Chassis adaptations; Engine control; Fuel/energy consumption optimization; Regulations (horse-power, period of use, mental state); Exhaust/emission control			④ ⑤ ⑥				① ② ③
VEHICLE NAVIGATION: Route planning; Drive by wire; Traffic optimization; Green light optimal speed advisory; Dynamic route changing and optimization; Position finding			④ ⑤				③
SAFETY SERVICES: Standalone Vehicle (ABS, ESP, ACC, BAS, EBA, SRS); Vehicle-2-Vehicle (Forward collision warning, Wrong way driving warning); Vehicle-2-Infrastructure (Automatic crash-alert)			④ ⑥				① ② ③
MAINTENANCE APPLICATIONS: Remote diagnostics; Vehicle firmware update; Just-in-time (JIT) repair notification service; Car/fleet management; Status information (cruising range, service intervals, etc.); Wear part observation	EXTENSIVELY RESEARCHED AND COVERED BY LITERATURE		④ **RESEARCH FOCUS**	ILL-SUITED AND APPLICABILITY DOUBTFUL		EXTEN-SIVELY RE-SEARCHED	③ **RESEARCH FOCUS**
ROAD PRICING TAX INSURANCE: Free-flow tolling; Drive-through payment; Automatic access control; Personal data synchr.; Fleet management; Personalized (distance-based) insurance policies; Cause-based taxation			④ ⑥				① ③
COMMUNI-CATION: Internet access; Answering E-Mails; Phoning; V2V assistence services; Emergency aid service; Remote diagnostics; Traffic information; Point of Interest notification			④ ⑤				① ③
ON BOARD ENTERTAINMENT: Car stereo control; MP3/DVD player; Game consoles; Media players; Satellite TV/radio; Voice/knob control; Portable device interfaces; Illuminating keys			④ ⑤				① ③

Fig. 5.2: Driver-Vehicle Interaction (DVI) modalities versus application domains.

5.4 Hypothesis I

Implicit Expression Decreases a Driver's Cognitive Load

Explicit Driver-2-Vehicle (D2V) interaction has been elaborately investigated and applied for most conventional car handling activities (such as steering, route-planning, changing gears, etc.), requiring active participation of the driving person and in turn leading to an increased cognitive workload.

It is hypothesized that the application of *implicit* input modalities in the emerging field of person-centric applications or services would offer great benefits for the driver, such as (*i*) decreased cognitive load in conjunction with the task, (*ii*) less distraction from the primary task of driving, or (*iii*) increased driving convenience.

Particularly, activities demanding continuous data input from the driver, such as identification or authorization services or applications based on activity recognition[22], would benefit from such an implicit input mode.

Up to now, personalized services in the car employing established forms of interaction are not at all or only infrequently used due to their complexity or the anticipated generation of an additional workload. Evidence that person-related properties and/or activities can be captured implicitly and processed in vehicles – and thus require no extra attention and generate no distraction – is assumed, but still missing to this day.

Subhypothesis I.i The sitting postures of all persons sharing a specific car are significantly diverse, so that that they can be employed for driver identification and accordingly enable personalized car services.

Subhypothesis I.ii Dynamically captured traces of posture patterns allow driving conditions to predicted prior to their occurrence.

Subhypothesis I.iii Sitting postures can be employed as a means of implicit vehicle control.

[22]In detail, this could include tasks like permit or deny engine start, apply specific engine settings (horsepower regulation), utilize personal settings for seat, mirrors, and air conditioning system, use a personalized car-insurance policy or adopt cause-based taxation, etc. to the point of "Fitness to Drive" applications like detecting a tired or drunk driver by real-time monitoring of his/her physical and mental behavior.

5.5 Hypothesis II

Vibro-tactile Notifications Improve a Driver's Response Time in the Driver-Vehicle Interaction Loop

Vehicle-2-Driver (V2D) feedback in automotive applications is mostly delivered with the sensory modalities *vision* and *hearing*, covering at least 80% to 90% of the total notification demand. With the steady rise in the number and complexity of infotainment devices, communication appliances, or Advanced Driver Assistance Systems (ADAS) and the plethora of information they provide, both the *visual* and the *auditory* feedback channels increasingly operate at full capacity. Possible consequences of these issues are (*i*) an impaired attention to the main task of driving or to other important notifications, and (*ii*) operation errors caused by distraction or high cognitive load.

Physically or environmentally induced restrictions influence both quality and quantity of notifactions via *visual* and *auditory* (and perhaps other) sensory channels. For instance, the eye would be affected by glaring or reflecting light, poor visibility in foggy situations or in bad weather, or by different kinds and behavior of receptors responsible for day and night vision, while auditory feedback could easily be missed and most driving situations are characterized by superimposed distracting noise originating from the engine, the environment, communication, or even the car stereo. Cell phone operation, as one specific example, has been shown to be responsible for affecting the driving performance as a function of all of these forms of distraction, namely (*i*) looking at the phone while dialing (visual), (*ii*) holding conversation with a person (auditory), and (*iii*) focusing on the conversation rather than on steering the vehicle (cognitive).

Apart from the *visual* and *auditory* notification channels, cognitive humanists specify the three exteroceptive senses *touch*, *smell* and *taste* to be incorporated in any Human-Computer Interaction (HCI) system. The contribution of the modalities *smell* and *taste* is quite unclear; however, the sense of *touch* offers great potential for improving Driver-Vehicle Interaction, but is underemployed today.

It is hypothesized that a Vehicle-2-Driver feedback system utilizing *touch* sensations discharges the load of the *visual* and *auditory* sensory channels, and consequently improves driving safety and comfort. The sense of *touch* as a non-cooperative[23] interaction channel seems to be qualified as a universal feedback medium in vehicles, and should be usable (*i*) cross-car (sports car, van, lorry), (*ii*) cross-seat (body-contoured seat, seat with air or oil suspension, "normal" seat), (*iii*) cross-person (any profession as well as any age or gender – for instance rookie, se-

[23]The expression "non-cooperative" in this context means that information is transferred from the driver to the vehicle even if the driver is neither aware of the transfer nor actively cooperating with the interface.

nior driver, professional championship driver), and (*iv*) cross-environment (gravel, tarmac or concrete paved road).

Findings concerning both the general applicability and/or performance of touch sensations in the automotive domain and evidence for the assumed relief of cognitive load when using the sense of touch are missing today.

Subhypothesis II.i The supplementary use of vibro-tactile feedback in Driver-Vehicle Interfaces relieves the visual and auditory sensory channels of cognitive load, and furthermore reduces the processing time required for the sum of the two motoric sub-processes perception and expression[24] of the Driver-Vehicle Interaction loop.

Subhypothesis II.ii The influence of age and gender[25] on stimulus perception and reaction times, which has been proven to affect all the sensory channels vision, hearing, and touch, is the lowest for the haptic sensory modality (and legitimates the extensive application of the sense of touch in future interfaces).

Subhypothesis II.iii Sex and age-related performance deterioration does not affect the order of the sensory modalities with respect to the response time.

Subhypothesis II.iv The sense of touch as an additional feedback channel in vehicles may replace driver-vehicle notifications delivered by the modalities vision and hearing.

[24] A detection of the perception time exclusively would be difficult; however, when employing similar methods for articulation, this portion can be easily deduced.

[25] Particularly the parameter age is of increasing interest as an overaging population and an increased expectation of life are assumed and at least partly proved.

Outline

The remainder of this work is structured as follows:

Chapter 6

"Vibro-Tactile Articulation and Presentation" (p. 41) starts with a short overview of information channels and their applicability in the automotive domain and proceeds with a motivation for using tactile displays in Driver-Vehicle Interaction (DVI). A special section is dedicated to a definition of haptics and related terms and a detailed study of research work in the field of vibro-tactile displays in Human-Machine Interaction (HMI).

Chapter 7

"The Driver as the Weak Point in Interaction" (p. 67) highlights factors leading to high cognitive load of the driver and suggests solutions for managing the workload.

Chapter 8

"Driver Activity and Notification Demands" (p. 71) considers questions regarding a driver's information demand, discusses driver distraction induced by erroneous notifications, and gives a suggestion for selecting appropriate sensory modalities for a certain notification demand.

Chapter 9

"Advanced Driver Assistance Systems (ADAS)" (p. 83) gives a short overview of state-of-the-art and emerging ADAS in order to get an assessment of possible operational areas in the scope of vibro-tactile Driver-Vehicle Interaction.

Chapter 10

"Vibro-Tactile Interfaces" (p. 85) begins with an examination of the human tactile system focusing on cutaneous mechanoreceptors, gives a selection of suitable technologies for stimulation of the skin, and finally describes the approach and features of a vibro-tactile alphabet. In this context, a notation for tactile patterns is introduced and a definition of commonly used terms (particularly "tactogram" or "LOA") is given.

Chapter 11

"Analytical Methods and Methodology" (p. 107) describes methods applied for the evaluation and interpretation of data recorded and preprocessed during the experimental phase.

Chapter 12

"Experiments and Evaluations" (p. 119) describes the goals, experimental settings, and the results of various experiments conducted within the scope of this work.

Chapter 13 – 17

"Predeterminations for Investigation" (p. 185), "Reflecting on the Hypotheses" (p. 186), "Experiments: Lessons Learned" (p. 192), "Conclusion" (p. 194), and "Future Prospects" (p. 195) provides a discussion of experimentally determined results, compares primary benefits and drawbacks of the chosen approach, and gives suggestions for further work in the area of implicit interaction in vehicles.

The **Appendix**

(p. 199) contains supplementary material, such as an overview of utilized sensor and actuator hardware, a section on biometric identification technologies, a part on physiological senses and proprioception, and others.

"Science cannot solve the ultimate mystery of nature. And that is because, in the last analysis, we ourselves are part of nature and therefore part of the mystery that we are trying to solve."

Max Planck (1858 – 1947)

German physicist and Nobel Prize Laureate (Physics)

Part II

Driver-Vehicle Interaction

6 Vibro-Tactile Articulation and Presentation

The emergence of new Driver Assistance Systems and Driver-2-Vehicle Interfaces brings with it an increased demand for interaction and communication on the one hand, and – in conjunction – the opportunity to change the style of interaction on the other hand. This creates, or even requires, the potential for establishing new modes of input and/or output. In order to avoid cognitive overload, and with consideration of the fact that *visual* and *auditory* sensory modalities are already highly stressed, it is recommended to utilize the remaining traditional senses *touch*, *smell*, and *taste* for covering additionally emerging information or supplementary notification demand.

Fig. 3.1 on p. 24 lists all sensory modalities which might play a role as an information source in vehicles. All of the items which are available in the car (and for whom a utilization makes sense), can be grouped into larger units – the five traditional modalities of sense. Text, graphics and animations belong to the *visual* sense, speech is the *auditory* sense, tactile and kinesthetic stimuli are a part of the sense of *touch*, *olfaction* and *gustation* are again self-contained senses.

Identified potential as well as limitations of the individual sensory categories are based on a "snapshot" of today's available technology and applications. Stated suggestions are therefore partly tentative, but could – with further technological advance – lead to interfaces which are not even conceivable today.

6.1 Excursus: Sensory Modalities

The following paragraphs give a short overview of the different (traditional) sensory modalities and their suitability for an application in Driver-Vehicle Interfaces. Moreover, they include a classification of the sense of *touch* into the entire sensory system and a short comparison of the pros and cons of the different channels. An overview of the second group of sensory modalities, the *physiological senses*, is presented in Appendix C.1: "Physiological Senses".

6.1.1 State of the Art in In-Car Interaction

In conventional user interfaces, and the automotive domain is considered to be in this category, interaction is very often done via the modalities *vision* and *hearing* only [171], while the remaining senses (*touch*, *smell*, and *taste*) are typically not considered at all. In general, this is an appropriate choice as it is a fact that information in HCI is primarily delivered via the visual and auditory senses ([25], [49], [164], [51], [64]).

As information is becoming more and more complex on the one hand (external scenarios such as the environment with all its conditions and situations can be cited as but one instance) and the number of appliances or services is steadily rising on the other [172], the demand for new forms of interaction and/or communication is growing. Effective interaction between drivers and their vehicles plays an ever important role, strengthened by the fact that none of the channels has unlimited capacity for processing and transferring information.

Changing a driver's behavior during his/her driving task would provide support for these demands. One possible solution, following the model of "Joint Cognitive Systems" (JCS)[26] has been presented with the so-called "Driver, Vehicle and Environment" (DVE) model, extensively considered, for example in [174], or [173, p. 17].

Vision

In current Driver-Vehicle Interfaces, the *visual* sensory modality is the dominant feedback channel; however, the *visual* attention very often needs to be focused on the road or the environment to observe changes and activities [175] (visual perception is an active process, resulting from an interaction among external inputs, expectations and other information processing features of the driver).

The visual system has been estimated to process up to 90% of all the information occurring in and around the car [75], [25], but is – as all other modalities – limited in its bandwidth. As a consequence, a visual interface should demand as little visual attention as possible in order for the driver to interact efficiently with it. This assumption has been substantiated by research on the visual workload, which has shown significantly reduced "eyes on the road time" when drivers interact with vision-centered in-vehicle systems [176].

Hearing

The *auditory* information channel, as the modality with the second highest impact ratio regarding the amount of information delivered (up to 15%, [50]), has been applied increasingly (e. g. in navigation or Park Distance Control (PDC) systems) in order to compensate for the disadvantages of the visual channel. Nevertheless, sound notifications often compete against other auditory sources, such as motor or environmental noise, conversation of passengers, or most prominently the car stereo. In addition, auditory feedback is characterized by the fact that it could easily be missed and is quickly considered relatively annoying (for instance, the tone signal indicating that the seatbelt is not fastened) [175].

[26]In HCI systems it is always easier to describe the functionality of the technological system (i. e. the machine) than to describe the functioning of the human operator. As a result, the interaction between humans and machines became the most important part to consider. Cognitive Systems Engineering (CSE) shifts the focus from the internal functions of either humans or machines to the external functions of the Joint Cognitive System (JCS). As a consequence, the boundaries (between system and environment and between the parts of the system) must be explicitly defined [173].

6.1.2 Potential of Smell and Taste

Beside *touch* – the key topic of this research work – the sensory modalities *smell* and *taste* potentially have the capacity to enhance interaction in Driver-Vehicle Interfaces and consequently, to relieve cognitive load from *vision* or *hearing*.

However, initial considerations discount the use of any of these two channels for in-car perception sensing. Gustation, for instance, is considered not applicable because no connection between *taste* and activities involved in operating a car has been identified [25, p. 78]. The same applies to the sense of *smell*, although some properties in the vehicular context are perceivable by the *olfactory* sense, e. g. environmental scents, or the smell of burning or leaking oil. Another issue related to the modalities smell and taste is the ever existing problem of the composition of odors and gustatory substances [111, p. 210].

Olfaction

Smell is a *chemical sense* and has attracted only little interest from the designers of multimodal systems until now. Therefore, the application of the *olfactory* sense has virtually been limited to research [111]. In addition, the impact of olfaction on tasks already using multimodal interfaces is still unclear, and thus has not been considered in this work. Nevertheless, the advantages and disadvantages of olfactory displays have been explored (list adapted from [112, pp. 113–117], [111]):

Advantages

(*i*) The *olfactory* sense is a peripheral medium, and thus does not require manipulation or eye contact. Eyes and ears-free use is of increasing importance, especially in situations in which the visual and auditory information channels are heavily loaded or occupied (like controlling a vehicle).

(*ii*) A constant aroma will become unnoticed over the course of exposure (but humans may still be aware of it upon conscious reflection). This would be an advantage in the automotive domain, because it does not cause distraction of the driver or decrease his/her free capacity of attention. Of particular importance is the perception of sudden changes in the "olfactory landscape" which can be guaranteed by the attribute "aggregation of smells" as indicated below.

(*iii*) Olfaction is, unlike vision or hearing, a hidden sense. Previous studies have shown that odors can influence alertness – this trait could be applied for controlling or stimulating a vehicle driver's vigilance.

(*iv*) An emitted scent will not instantly disappear (as a burst of music or a flash of light), but fade out. This is a benefit because visual information in vehicles often is overseen and auditory notifications are missed.

(*v*) Smells can be aggregated. Unlike auditory information, where overlayed channels of voices or sounds often are misconceived, odors can be combined and added. This allows one scent to be simultaneously compared to another.

(*vi*) Scents are dispersed. As opposed to visual notifications, where the driver needs to turn in a particular direction to perceive the information, odors are spread everywhere. Moreover, humans have the ability to differentiate among scents located at various positions.

(*vii*) Scents have an "alerting quality". The sudden recognition of odors like gas or burning is a learned, not an instinctive, response. This could be used in cars for rapidly detecting engine failures, and moreover could be adapted for future displays where particular scents are standardized for delivering important messages.

Disadvantages and Problems

The application of the olfactory sense as data carrier is subject to a number of limitiatons and restrictions.

(*i*) Olfactory displays have a low resolution. Humans can only distinguish between a limited number of scents, and it is very hard to determine absolute concentration levels in diluted scents. Furthermore, odors have (due to their chemical characteristics) a low spatial and temporal resolution.

(*ii*) Odors generate crossfeed interference. Emitted smells disperse only slowly and thus may interfere with scents appearing later. It cannot be ensured that two or more scents will not interfere with each other.

(*iii*) Odors have the potential to provoke allergic reactions and bad odors will annoy people. Furthermore, the perception of odors is not consistent for all people.

(*iv*) A scent produced and emitted at a specific location for one individual will spread and can be sensed by others (this property excludes the use of scents for exact localization).

(*v*) If a designated person misses a scent, it is gone (there is no intrinsic history).

(*vi*) It is still a large problem to generate arbitrary scents on demand and from a limited number of primary odors.

Gustation

Gustation (or taste) is a form of direct chemoreception[27] and is another *chemical sense* which is often confused with flavor[28]. It is a sensory function of the central nervous system and the receptor cells for taste in humans are primarily found on the surface of the tongue. Taste refers to the ability to detect the flavor of substances such as food and poisons. Classical taste sensations include sweet, salty, sour, and bitter [110]. According to Small [178], the primary function of gustation is to identify substances in food and drink that may promote or disrupt homeostasis.

Considering the characteristics of gustation, a significance for utilizing the gustatory sensory channel in the automotive domain today is generally not given. As there are no obvious fields of application, this modality is not studied further in this work.

6.1.3 Summary

The synopsis presented here clearly indicates that *vision* and *hearing* are still the two most widely applied modalities (not only in the vehicular domain), but increasingly factors such as cognitive overload, distraction (caused by excessive information), or physical limitations (light conditions, noisy environment, etc.) affect the interaction quality, and lead to misinterpretations, danger situations, accidents, and inconvenience. On the other hand, the sensory modalities *smell* and *taste* contribute only to a minor degree to interactions in vehicular applications today (or are even unsuitable in the case of taste) – thus, an estimation of their potential is not too easy.

Increased attention is dedicated to the utilization of the sense of *touch* as an additional sensory modality for driver-vehicle ("input") or vehicle-driver ("output") interaction. Haptics has already shown its potential in diverse fields of application, e. g. traditionally for assisting the blind and visually impaired, or in interactive functions while driving a car [175]. Furthermore, as stated by Wicken in his Multiple Resource Theory (MRT) [59], humans can perceive information concurrently when using different modalities. Applying this theory to the automotive domain creates the potential for drivers to perceive a larger quantity of information without increasing their level of attention (LOA) by utilizing the sense of *touch*.

Another fact influencing the selection of a certain sensory modality (in this work *vision*, *sound*, and *touch*) is its dependency on the interaction task for a specific situation. According to Schäfer *et al.* [179], or a guideline from the European Telecommunications Standards Institute (ETSI) [1, p. 19], this implies the selection of a device together with one or more suitable interaction modalities for the best possible interaction performance/quality.

---

[27] Chemical stimuli activate the chemoreceptors responsible for gustatory perception.

[28] As opposed to taste, flavor is a fusion of multiple senses. To perceive flavor, the brain interprets not only gustatory (taste) stimuli, but also olfactory (smell) stimuli and tactile and thermal sensations [177].

6.2 Motivation for Tactile Displays

Apart from the designated usage of the sense of touch in vehicles, a wide range of application domains using tactile information have emerged in the last years. This is another indicator demonstrating the importance of the sense of touch for future interfaces.

Among others, the sense of touch has been successfully applied to resolve spatial disorientation of pilots, to convey an aircraft's position and motion information, to support orientation awareness of astronauts, to assist doctors in teleoperation, to provide information from inside the patient's body to a surgeon's fingertips during robotic surgical procedures, and to provide route-finding applications for pedestrians [13, p. 658].

6.2.1 Complex Vehicle Handling

Due to technological improvements, drivers are forced to access more and more assistance systems and services in the car; for instance, systems for route navigation, traffic information, entertainment, mobile communication, etc. have been integrated. Aside from the increasing number and complexity of these systems, the car itself provides more and more information too, from informative (tire pressure, fuel cruising range, parking guidance) to warning or even danger messages (speed limit warning[29], road work ahead, traffic jams, accidents, high cooling water temperature).

The consequences of interacting with these information systems in vehicles using traditional forms of interaction can be negative, and furthermore would affect the attention of the driver to his/her primary driving task. Subsequently, this leads to high cognitive load or even overload, and in succession results in longer response times and decreased road safety. Overload in this context means that the driver is unable to process all relevant information necessary to perform the primary task of driving or harmlessly operate the car [165]. As one feasible solution, the application of the sense of touch as an accessory sensory channel should be responsible for compensating for the high level of cognitive load.

6.2.2 Haptics as Support for Visual/Auditory Interaction

Until now, Driver-Vehicle Interaction (DVI) has used the *visual* and *auditory* sensory channels exclusively for transmitting information to the driving person (individually or combined). One of the most important advantages of using vibro-tactile displays would be that they might not

[29] In 2008, automobile manufacturer Opel presented a system called "Opel Eye" [180], (extra equipment for the new "Insignia" model) which can identify round road signs and notifies the driver about speed limits or prohibited overtaking zones), http://www.opel.at/page.asp?id=2005090320284932IM7, published online June 18, 2008.

be affected by the following limitations, occurring during information processing through the visual and auditory sensory modalities:

Cardullo [181] reported that vision is affected in high-acceleration environments, and that the result is a loss of visual acuity – which leads directly to performance deterioration. Van Veen and Van Erp [182] found that tactile presentation of information (using vibro-tactile stimulation on the torso) in high-G load environments is much better suited than using the visual channel (and is hardly impaired up to a force of 6-G). Moreover, the sense of touch is insensitive to changing conditions of illumination (for instance, day or night vision[30], passing through road tunnels, glaring sunlight, rain, snowfall or fog, etc.), and is not reduced to the current direction of fixation in its performance. Furthermore, it is not adversely affected by background noise or overlapping of several audio sources like voices, etc. (high levels of noise may be associated with higher accident rates [185], tactile stimuli can be used as a sensory aid to compensate for the overload of the usual means of communication, as shown in certain military operations [186]). The utilization of the sense of touch is rather uncommon today [4], although it has considerable potential: The tactile channel is (*i*) always ready to receive information or pay attention, (*ii*) a private medium (nobody else would realize notifications), and – as the experiments have validated – (*iii*) useable in a natural, intuitive, and non-distracting manner [168].

Early studies of Geldard [187] or Von Haller Gilmer [188] pointed out that touch sensation may have an automatic ability to capture attention. In addition, a tactile body display may have the ability to present 3D spatial information in a convenient manner, while it has to be transferred to a 2D representation when using a traditional visual display (Van Veen *et al.* [182, pp. 175], Jeram [189], Jeram and Prasad [190]).

Van Erp and Verschoor [191] investigated tracking performance of vision and/or touch sensations. They found that a tactile display performs worse compared to a visual display (probably due to the larger resolution of the visual); but on the other hand, the tactile modality is better suited to present information that does not contain external disturbances.

Using tactile stimulation instead of visual or auditory stimuli has the ability to make driving safer by reducing the workload of the latter two channels and furthermore supports the "hands on the steering wheel" and "eyes on the road" strategies [168, p. 3].

[30]The human retina contains two types of light-sensing cells, (*i*) cones (responsible for sensing and discriminating colors; they do not sense very dim light, thus they are used for daytime vision), and (*ii*) rods (liable for peripheral vision in bright daylight, in very dim light, and in mixed lighting conditions such as on a dark road at night with bright oncoming headlights; they only see in black/white or gray). In normal conditions, the cones function with and complement the functions of the rods [183]. Capanzana [184] stated that 10.5% of all mothers are affected by night blindness – at least they are unable to drive safely or comfortably at night.

Tactile Information in Driver-Vehicle Interaction	
ADVANTAGES	DISADVANTAGES
Potential to lower the level of attention (LOA) required for visual and auditive channels of perception.	The skin can get adapted to permanent stimulation.
Potential to lower the cognitive load, because tactile displays present information implicitly (or more intuitively than auditive/visual displays).	People are not used to tactile displays today. Tactile information presentation would require training or learning.
Potential to draw permanent and directed attention. The skin is always ready to receive tactile information - this is a plain advantage compared to the visual channel. If we don't look at the visual display, we will not receive information. Tactile information is always perceived, the observer does not have to make eye or head movements.	Mechanical stimulation can interfere with other tasks - a strongly vibrating steering wheel could influence steering precision.
Stimulus locations on the body are directly mapped in a self-centered reference frame - this makes the skin an interesting information channel for requiring an egocentric view.	Tactile display technology is very limited in its resolution (compared e.g. to the visual display technology, which offers resolutions of millions of pixels per image).
A tactile display allows distal attribution - extends bodily experience beyond the physical limits of our torso (stimulus on the skin is attributed to an event or object in the outside world).	The most spatially sensitive area is on the fingertips, but displays on the fingertips are mostly impractical (they are needed for controlling devices, holding tools, etc.)
	Tactile displays have to be in contact with the person (the skin) - the design and placement of tactile interfaces are therefore strictly limited.

Fig. 6.1: Advantages and drawbacks of using the tactile sensory channel for interaction in vehicles (adapted from Erp [3, p. 23]).

6.3 Definition of Terms

For the terms *haptics*, *tactile* or *tactition*, *somesthesis* and related notions, multitude of definitions can be found in the literature. Those of Tan and Pentland [192, p. 581], or [193, p. 85] seem to be well founded and therefore should be used as a starting point for further investigations in this research work.

Somesthesis (or somatic sensibility, Gross *et al.* [194]) refer to (*i*) the primary qualities of skin sensations (touch pressure, form, texture, and vibration), which are commonly referred to as the *tactile* or *cutaneous* sense, and (*ii*) more complex sensory experiences (senses of spatial pattern, contour, shape, and the senses of movement and position of our limbs) which are identified as *kinesthesis* or *proprioception* [195, p. 9], and belong to the self-contained sensory modality proprioception (for a clarification of the interrelationship among *kinesthesis*, *proprioception* and related terms as well as for the definition of terms for their usage in this work see Appendix C.2: "Proprioception" on p. 217).

According to Hong Z. Tan [4] and others (e. g. Craig *et al.* [196, p. 314], Thangavelu [197], or Meilgaard *et al.* [36]) the property *haptic* (or *haptics*, *tactition*) refers to (*i*) sensing or perception (*somesthesis*, limb position and movement) and (*ii*) manipulation[31] through the sense of touch (see Fig. 6.2), which means "the ability to perceive the environment through active exploration" [130]. *Cutaneous* or *tactile* sensing refers to the (passive) awareness of stimulation of the outer surface of the body while *kinesthetic* sensing refers to an awareness of limb position and movement as well as of muscle tension [4].

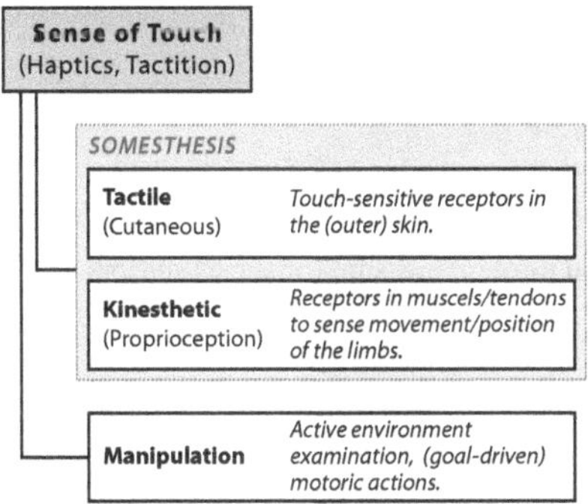

Fig. 6.2: Classification of the sense of touch.

The two physiological sensory modalities *pain* and *temperature* would offer at least limited potential for an application in vehicles, but they are not considered in this work as they (*i*) can be strictly separated from *cutaneous* and *kinesthestic* senses[32], and (*ii*) are understand to be fully independent. For a detailed description of pain and temperature see the Appendix "Physiological Senses" on p. 214.

6.3.1 Force Displays

Touch-sensitive devices can be seen as *force displays*. In the context of this work, the term "display" (or "notifying" as indicated in Fig. 6.3) refers to that part of the HCI or DVI system responsible for transferring information from the vehicle to the driver. Contrary to this, the

[31] Haptic interaction is bidirectional. This is different from visual and auditory sensory modalities which are only input systems. A typical example using both directions, sensing and manipulation, would be the reading of braille text by the blind [4].

[32] Webster's Online Dictionary: Definition of "tactition", http://www.websters-online-dictionary.org/definition/feel, last retrieved August 31, 2009.

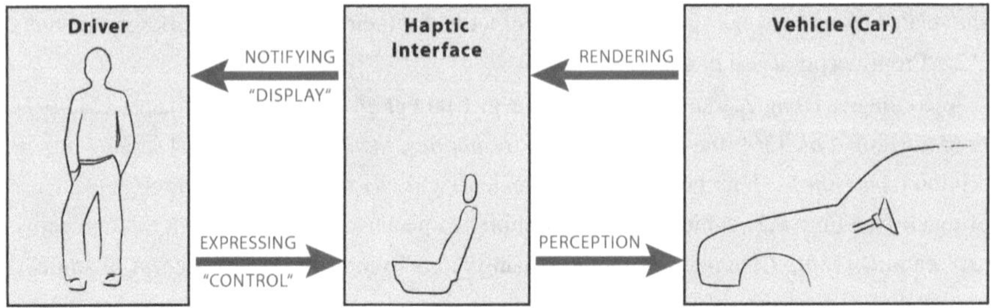

Fig. 6.3: Information flow in vehicular interfaces using touch sensation and manipulation (adapted from Tan [4]).

term "control" (or "expressing") refers to a D2V interface that is used by the driving person to control certain processes of the vehicle (see Figures 1.1 and 6.3). The computer or vehicular system would, depending on the direction of information flow, either render a haptic world, or perceive haptic information through pressure (or shear) sensors.

6.3.2 Summary

In the research community dealing with *haptics*, *tactition*, or the sense of *touch*, the property *haptic* is increasingly used synonymously for all the terms referring to sub-capabilities only (such as, for instance, *cutaneous*, *kinesthesis*, or *proprioception*) and, for that matter, is used in the same manner in this work. A more sophisticated investigation into the sense of *touch* is presented in the scientific works by Robles-De-La-Torre, [130, pp. 27] or Fryer [195].

Appendix C.2 "Proprioception" summarizes an elaborate examination of issues related to kinesthesis and the sense of touch, underpinned with statements of experts in this research domain (scientists, psychologists), with the aim of defining a common base for usage and interrelationship of the terms.

6.4 The Skin as a Sensory Organ

The skin is the body's largest organ, is sensitive to pressure, vibration, position and movement, temperature, pain, electric voltage, etc., and serves several needs. One specific function of the skin is to "[..] act as an external sensory organ having several kinds of sensors" [198]. Specific to the sense of *touch* is that it is a "hidden" sense or peripheral medium, operating eyes and ears-free, and moreover providing information even to visually and hearing impaired individuals.

Beside other sensory organs, the skin includes mechanoreceptors, responsible for pressure, vibration and slip [199]. The (four) different kinds of mechanoreceptors (Ruffini corpuscles, Merkel discs, Meissner corpuscles, and Pacinian corpuscles) are embedded in the skin tissue, each at a specific depth and with more or less appropriateness for stimulation via a vibro-tactile display. The most commonly exploited mechanoreceptors for vibration sensation in tactile display applications are the Pacinian corpuscles (PC) [200]. It should be noted that tactile perception results from the sum of the afferent input from Pacinian corpuscles as well as the unconsidered three other types of receptors [195].

The selection of the best qualified sensory organ was oriented towards the general requirements (i) spatial resolution, (ii) sensitivity, (iii) distribution/availability over the body, and (iv) adaptation speed. According to the characteristics of the individual cutaneous mechanoreceptors and with regard to the requested demands, the Pacinian corpuscles (PC) have been chosen because they (i) provide rapid adaptation, (ii) are embedded in almost all regions of the skin, (iii) have an acceptably high resolution, and (iv) are suited best for (high-frequency) vibration exploitation.

The relevant parts of the skin and their characteristics are considered in more detail in the Section "Cutaneous Mechanoreceptors" (p. 89).

6.4.1 Stimulation Technologies

A large number of techniques are available for generating and transferring vibrations to the driver, for instance, pneumatic valves, piezo-electric crystals, nitinol, electro-tactile actuators, electro-mechanical tactors, pneumatic or thermal actuators, etc. For a detailed overview and a brief technical description of the different technologies, see "Types of Stimuli" on p. 87.

The technology used in this research work has been selected as a function of the demands for in-vehicle utilization and the interaction with humans. Based on the requirements (i) strong forces (the capability of transmit sensations through clothing), (ii) low operation voltage, (iii) price, (iv) distraction level (e. g. deflating air in pneumatic devices), and (v) bandwidth, vibro-tactile actuators have been identified as being best suited.

6.5 Research on Vibro-Tactile Stimulation

In recent years, researchers have investigated the potential of vibro-tactile displays in human-machine interfaces. Because of the human skin's capability to sense vibrations virtually everywhere on the body, applicability and performance of tactile information processing has been studied separately for interfaces applied to specific body regions.

In this research work, the focus is on the vehicle seat, thus the center of interest is findings from bottom and back displays.

6.5.1 Waistbelt (Torso)

Jones *et al.* [201] tested a wirelessly controlled tactile display (built up from a 4 by 4 array of vibrating motors, and mounted on the lower back) for assisting navigational tasks. The evaluation of the vibro-tactile display regarding outdoor navigation confirmed that the interpretation of presented vibration patterns for directional or instructional cues performs almost perfectly. A further experiment showed that the ability to recognize vibro-tactile stimulation on the back was superior as compared to the forearm.

Lindeman *et al.* [202] designed the "TactaBoard" system to individually control tactor systems. In the conducted experiment, a 3 by 3 array of vibrating elements was used to investigate visual and vibro-tactile stimulations. The researchers reported that the performance increase was most significantly enhanced by visual cueing. Haptic cues alone (unimodal) provided a (lower) significant performance increase too. Considering these results, they suggested using the vibro-tactile cue when visual cueing is not practical.

Jan van Erp [203] investigated the direction in the horizontal plane using a linear tactor display with fifteen vibro-tactile elements placed on the participant's torso (with the oscillating frequency fixed to $250Hz$). His results confirm that people have the ability to indicate an external direction that matches a vibro-tactile point stimulus on the torso.

Van Veen *et al.* [204] have investigated whether a tactile torso display can help to compensate for degraded visual information. A tactile display, consisting of up to 128 vibrating elements oscillating at $160Hz$, arranged in twelve columns and five rows, and attached to a vest, was developed. Results showed an increase in operation performance for tactile display variants compared to operation without a tactile display.

Lindeman *et al.* [205], [206] have developed a body-worn garment to deliver haptic cues to the whole body. Initial evaluations regarding "haptic" navigation for collision avoidance or obstacle detection in real or virtual worlds have been presented.

Elliott *et al.* [207] investigated tactile waistbelts for the transmission of navigational information in order to leave eyes and hands free for other tasks. The presented tactile display was expected to improve the high visual and cognitive workload of persons by reason of Wicken's Multiple Resource Theory. Experimental results showed that navigation with the tactile display reached an accuracy of 100.00%. In comparison, in navigation with a GPS device, 95.60% accuracy was reached, and 86.70% for navigation with an ordinary compass. A similar study was conducted by Duistermaat *et al.* [208]. They evaluated the performance of navigation and target detection tasks in night operation, again with a vibro-tactile waistbelt, a GPS device, and a head-up navigation device and found that in all cases the tactile system was rated higher than the GPS system, and in most cases higher or as high as the head-up navigation system.

Tsukada and Yasumara [209] introduced a wearable tactile display called "ActiveBelt" for transmitting directional cues and found that (*i*) the vibration on the back is felt weakly compared to vibrations on the abdomen, (*ii*) it is more desirable to activate the vibration elements only when users get lost in their navigation task rather than activating them constantly, and (*iii*) for "walking" navigation the disposition of four tactors would be enough (while in the experimental setting eight vibration elements had been used).

Erp *et al.* [210] investigated the feasibility of navigation information presented on a tactile display instead of using a common visual display. Experimental results of the studies with a vibro-tactile waistbelt indicated that both types of display perform with similar reaction times (after a short familiarization phase).

Ferscha *et al.* [211] presented a body-worn, belt-like vibro-tactile notification system for assisting in raising humans' attention. The usability of the proposed system has been demonstrated in the field of worker safety in construction and industrial manufacturing. Evaluation results are encouraging for a new class of space awareness solutions based on vibro-tactile notification.

General Conclusions

The results of the diverse experiments with haptic waistbelts are interesting for vehicular applications too, e. g. for services benefiting from directional information, such as a traditional navigation system or a display showing Points of Interest POI[33].

6.5.2 Bottom and Back

Tan *et al.* [193] combined input from pressure sensors on the seat with output via tactor elements embedded in the back of the seat and integrated this system into a driving simulator. They analyzed patterns on the pressure array mat to determine whenever the driver intends to change lanes, and then gave attentional vibro-tactile cues. Furthermore, Tan and his colleagues described a two-dimensional vest-based system for mapping information about the surrounding traffic conditions onto the back of the driver and delivering warnings and guidance signals for navigation [212], [192].

Ertan *et al.* [213] introduced a haptic navigation guidance system integrated into the back of a vest. Experimental results confirmed that using a haptic display for navigation guidance would be a feasible solution.

Van Erp and Van Veen [168] processed navigation experiments triggered by visual, haptic and combined stimuli. Their tactile display consists of two strips of four tactors each, integrated

[33] A directional haptic display, embedded, for instance in the steering wheel, could notify about interesting points along the way like "[..] look at the monument on the front left" by activating the corresponding tactor element(s) and under consideration of steering wheel position and angle.

into the seat and vibrating constantly at $250Hz$. The remaining distance to a required course change was notified with a (decreasing) interstimulus interval. Findings indicated that a tactile navigation display results in faster reactions and lower workload and mental effort in association with the navigation messages than a visual display. They conclude their work with the statement "[..] the major promise of vibro-tactile displays is to improve the safety, efficiency, and fun of car driving".

Tan *et al.* [214] presented a 3 by 3 vibro-tactile array, embedded in the back of a seat, for displaying attentional and two-dimensional directional information to test participants. They found that valid haptic cues decrease the reaction time for the detection of a visual change, and that invalid haptic cues increase the reaction time (to a lesser degree) for the same task.

Van Erp *et al.*[132], [215] measured workload and reaction times for visual, tactile (eight vibrating elements integrated into the driver's seat), and multimodal navigation displays in a driving simulator. Their results demonstrated that navigation with a vibro-tactile display reduces the workload of the driver compared to visual displays (particularly in tasks with high workload). The fastest reaction thereby was achieved when using the multimodal display.

Ho *et al.* [216] investigated the potential use of vibro-tactile signals to present spatial information to car drivers (notification through vibro-tactile stimuli on their front or back). They found that the presentation of vibro-tactile stimuli can lead to a shift of visual attention that facilitates time-critical responses to visual events seen in distal space. Furthermore, they suggested that the crossmodal links in spatial attention between touch and vision may have a number of real-world applications in the future design of multisensory interfaces.

Ho, Tan and Spence [217] presented a discrimination experiment following the presentation of either vibro-tactile (two tactor elements, one on the front, middle of the stomach and another one in the middle of the participant's back) or auditory signals (two loudspeaker cones, placed near the front and the rear of the participant). Their experimental results confirmed that the use of vibro-tactile and auditory warning signals in automobile interface design to inform drivers of the occurrence of time-critical events outside the car is both effective and practical. Furthermore, their findings imply that response compatibility is an important factor in multimodal interface design.

Tan, Slivovsky and Pentland [192] described a static posture classification system based on a sensing chair that monitors posture patterns on the seat and backrest in real-time. They used a Principal Component Analysis (PCA) technique for static posture classification from pressure maps and reached a classification accuracy of 96% (for subjects known to the system) and/or 79% (for persons new to the system).

Cholewiak *et al.* [218] reported that many airplane accidents are due to a pilot's loss of spatial orientation during complicated maneuvers. They proposed a vibro-tactile vest, worn by the pilot, that indicates, for example, the direction of "up" on the chest and the bottom, and mentioned that such a system has the potential to save lives.

General Conclusions Vibro-tactile displays embedded in the seat seem to offer a valuable answer to the demand of the automotive industry to improve the Driver-Vehicle Interface in cars. Futhermore, employing the tactile channel in vehicles may relieve visual and auditory sensory channels of their high load, and therefore potentially provide a major safety enhancement. In addition, it will be important in future research to examine the quantity and quality of information that can be communicated via touch. Such information transfer can be supported by varying the intensity of the tactile stimuli or by presenting various patterns of stimulation ("tactograms") to represent different signals[34].

6.5.3 Head

Gilliland *et al.* [222] conducted a series of studies to explore the use of tactile stimulation of the human head to inform a pilot of designated situations in a flight environment. The study of localization performance revealed an accuracy between 93% (for 6 different stimulus spots) and 47% (for 12 spots) across the parietal meridian of the head. The studies on the performance of a localization task demonstrated that a tactile information display could be an integral contributor to improved situation awareness, but not without cost to other task performance.

Yano *et al.* [223] developed a vibrator suit to provide haptic sensation all over the human body, e. g. on forehead, elbows, knees and thighs. They used specific vibration patterns to let the user feel virtual objects, and in addition suggested their utilization for specific navigational tasks.

6.5.4 Feet

In the "Generic Intelligent Driver Support" (GIDS) research project, an active gas pedal (providing a counterforce when a speed limit is exceeded) was investigated in several situations, particularly in Collision Avoidance Systems (CAS) ([164], Verwey *et al.*). Janssen *et al.* [224] compared the performance of Collision Avoidance Systems with different system warning actions (namely via a buzzer, a red warning light on the dashboard, or an active gas pedal) and found that the haptic system performs best. In the project "Application of a Real-time Intelligent Aid for Driving and Navigation Enhancement" (ADRIADNE), Janssen and Thomas [225] evaluated different CAS under various visibility conditions (daytime, darkness, and fog) again

[34] People seem to have no difficulty in localizing vibro-tactile stimuli presented to their torso [203]. But it has been shown that participants (who have little prior experience of vibro-tactile displays) cannot count correctly if more than four vibro-tactile stimuli were presented at any one time (e. g. in [219], [220], [221]).

with a haptic gas pedal and other notification channels. Tests with different Collision Avoidance Systems evidenced that a CAS with a 4 second Time-to-Collision-(TTC)-criterion, in combination with a system action to the driver via an active gas pedal, reduces the percentage of small headways considerably without having contra-productive effects on other behavioral measures.

Godthelp and Schumann [226] have investigated haptic gas pedals too – they suggest that active control information may serve as an efficient information system.

Van Winsum [227] used tactile stimulation on the acceleration pedal to notify when a certain speed limit was reached (or exceeded). He evaluated this haptic display against an auditory display (message) and found that the tactile display resulted in decreased workload and in faster responses [168].

Kume *et al.* [228] studied tactile sensations by multiple vibrations on soles to minimize discomfort caused by wearing special devices. They integrated two vibro-tactile elements into the sole of each shoe and made use of phantom sensations elicited by these tactors. In their experiments they found that characteristics like location, movement or rotation could be perceived.

Frey [229] presented an alternative concept, not situated in the vehicular domain, for navigation using vibro-tactile information transmitted via the foot. The interface was motivated by the fact that visual and auditory channels of information are often highly loaded or required for other tasks, and that they cannot be used satisfyingly for navigation in these cases. The interface is intuitively understandable and thus, results in easy to use and feasible experience. It would be fairly easy to transfer the presented concept of haptic navigation assistance into the vehicle by enabling the pedals with vibration functionality or using the car seat instead of the pedals or shoes.

Kim *et al.* [230] presented a vibro-tactile display focusing on vehicular applications. The system consists of an array of 5 by 5 tactor elements which is mounted on the top of the foot for providing safety information to vehicle drivers. Feasibility tests showed a recognition rate of 87% for characters and 84% for patterns. Moreover, training improved the recognition rate significantly, which implies that after long term usage of such a device, the recognition rate will be further improved.

General Conclusions

Experiments with vibro-tacile feedback on the foot or in shoes confirmed that they can be applied to real situations. Such a vibro-tactile display could be used for supporting drivers in tasks like directional cues for navigation, lane detection or obstacle warning. It is important to note that (*i*) training on the symbols of the vibro-tactile alphabet increases the performance significantly, and (*ii*) the design or definition of appropriate, non-confusing stimulus patterns for providing information to the driver is absolutely essential for the later efficiency.

6.5.5 Hands

Burke *et al.* [231] performed studies on single-dimensional tracking tasks, executed simultaneously by the right and the left hand. The tracking was thereby performed with a visual display by the right hand (secondary task), and either a visual or tactile (kinesthetic-tactual) display by the left hand (primary task). They measured a significant increase in task performance for the secondary (visual) task in a dual-task situation, when using the tactile display for the primary task, compared to the situation with two visual displays.

Amemiya *et al.* [232] designed and evaluated a novel handheld haptic force display. The proposed system is based on the nonlinear characteristics of human tactile perception – the haptic display slips over the skin on the hand in one direction with a force proportional to acceleration. Their experimental results indicate that the frequency of acceleration change is more important in establishing the perception of the vector than the amplitude of acceleration (they stimulated rapidly adapting (RA) Meissner corpuscles).

Schumann *et al.* [233] investigated vibro-tactile feedback in the steering wheel and likewise found that haptic cues are better suited to inform the driver about a certain situation than auditory warnings.

Vitense *et al.* [234], [235] investigated multiple forms of feedback, namely auditory, haptic and visual, to evaluate how uni-, bi-, or trimodal feedback affects the performance of users. They found that user interfaces incorporating haptic feedback unimodally or bimodally with visual feedback would be more effective. Overall results substantiate the claim that multimodal feedback enhances user performance.

6.5.6 Buttocks

Lee, Hoffman and Hayes [236] investigated graded haptic warnings in vehicles, acting as a substitute for single-stage auditory warnings. For this purpose, vibro-tactile actuators were placed on the front edge and in the thigh bolsters of the driver's seat. Experiments showed that haptic warnings were interpreted more precisely than the auditory warnings (fewer inappropriate responses, lower level of annoyance, etc.), suggesting that haptic displays may be a viable alternative to auditory feedback sytems.

Sklar and Sarter [237] studied the parallel processing of large amounts of data distributed across visual and tactile (tactors around the wrists) sensory modalities in a complex cockpit system. Beside approved results (higher detection rates and faster response times for the haptic modality), they observed that tactile feedback (*i*) did not interfere with the performance of concurrent visual tasks, and (*ii*) was not affected in its effectiveness by concurrent visual tasks.

6.5.7 Arms

Kohli *et al.* [238] explored the use of tactile motion at different speeds for displaying information using a vibro-tactile system of three rings of five tactors each mounted on the upper arm. Their results indicated that test participants have little trouble with pattern identification (they achieved identification rates between 94% and 100%), but found absolute speed recognition difficulties (in the system used at least two speeds were distinguishable; the researchers stated that with further tuning and more training three speeds would be identifiable).

6.6 Touch-Based Driver-Vehicle Input

Apart from vibro-tactile stimulation (*output*) the *input side* also offers potential (applications based on sitting posture evaluation) and has to be mentioned. Touch-based input could be acquired e. g. by use of pressure sensor arrays; however, an overview of potential applicability is of secondary importance in this work due to the availability of related research papers dealing with the use of pressure sensors in office or wheel chairs:

(*i*) **Chair-movement recordings:** Fenety *et al.* [239]

(*ii*) **The chair as user interface:** Slivosky and Tan [240] [241], Tan *et al.* [192], Overbeeke *et al.* [242], Mota and Picard [243]

(*iii*) **Posture recognition in chairs:** Andreoni [244], Miller [245], Healey and Picard [246], Zhu *et al.* [247], Ishiyama *et al.* [248]

Further information is given directly in the corresponding studies in the experiment section. This short overview shows that pressure sensing in the automotive domain (for identification tasks or posture-based control activities) is a novelty and has not been the subject of much research effort so far.

6.7 Haptic Interaction in Vehicles

Touch-based interaction offers potential benefits when used in the vehicular domain. This is because most driving situations are affected by environmental noise or other distractions (for a detailed explanation see the Section "Notification-Induced Driver Distraction" on p. 72). Moreover, drivers are often required to multiplex their visual, auditory, and cognitive attention between the environment and the information device. Additionally, it has to be considered that in some contexts certain feedback options are not appropriate – as ringing cell phones are a taboo in quiet meetings, information-rich displays would not be permitted while driving [43].

The ever increasing functionality-richness of DAS has to compete with limited display capabilities and input paradigms. Considering small displays (e. g. the first release of BMW's "iDrive" control), the amount of information that can be presented on such a device is often organized in deep hierarchies – which creates additional navigational challenges for the driver.

Another current challenge for interface designers in these highly interactive domain is to transfer relevant information from the assistance system to the user in a digestible form. Haptic force feedback can be utilized to this end. The many possible approaches to devising intuitive haptic control-sharing cues can be differentiated by the degree of control retained by the user for a specific action (from a system, behaving completely autonomously on the one end, to a user totally responsible for interface control on the other). For instance, haptic cues offer a unique opportunity for navigational use in that guidance cues can be directly superimposed onto the physical control channel. This potentially allows for very immediate and easily integrated feedback to the user, but might also obscure the user's perception of the system if the user interface is not designed appropriately – and would then demand additional attention from the same [249].

6.7.1 Driving Dynamics

After detailed consideration, a vibro-tactile feedback system embedded in the vehicle seat seems to be a suitable platform for delivering haptic stimuli to the driving person. However, during acceleration and deceleration phases of the vehicle (e. g. at crossings, during overtaking or in sharp bends), passengers are subjected to a multitude of forces and vibrations (resulting from motor movements or from lateral forces released by steering the vehicle). Additional forces, even during "normal" steering, emanate from the roadbed especially when driving on gravel roads.

The impact of these forces has been investigated by Brewster and his colleagues [250]. They found that it is particularly difficult to interact with (small-sized or) mobile devices on buses or trains, where the journey can be very bumpy by reason of road or rail condition. Similar interaction modalities appear in vehicles where the driver wants to control technical appliances like a navigation system or an on-board computer during the ride. Furthermore, Brewster *et al.* [250] determined that there is only little evidence that tactile displays are beneficial in practical mobile situations because of non-expressive environmental influence (the underground already generates a baseline of vibrations which would interfere with haptic notifications).

These concerns can, however, be mitigated when intending to use touch signals for feedback, because it is a fact that vibro-tactile stimulation would be (*i*) much stronger than the permanent vibrations resulting from the environment, (*ii*) considerably different in its intensity than underground vibrations, (*iii*) perceivable all the time, even when other feedback modalities fail, e. g. audio notifications while driving at high speed (loud ambient noise), talking on the cell phone or communicating with passengers, or visual stimulation in driving situations where eyes are heavily charged, on changing light situations or dazzling sunlight, (*iv*) universally applicable, because touch receptors can be found all over the human body – therefore it would be possible to find locations for tactor placement in the seat, covered by drivers of any figure, or even places resistant against environmental interference.

6.7.2 Vibro-Tactile Stimulation: State of the Art in the Automotive Domain?

Numerous literature reviews regarding the utilization of the sense of touch or vibro-tactile interfaces have recently been presented, for instance by Gallace *et al.* [220] or Tan *et al.* [251]. Lee *et al.* [252] studied tactile interfaces for drivers to overcome some of the shortcomings associated with the use of the traditional visual and auditory interfaces.

A summary of their research, along with relevant results of others, is indicated in the work of Lee and her colleagues. The importance and benefit of applying vibro-tactile interfaces in the automotive domain is supported by several activities and observations about upcoming applications all around the world, as for instance:

(*i*) An early commercial input device, trying to resolve issues of interface operation by using haptic feedback delivered via a knob, was introduced as the "iDrive" control by automotive manufacturer BMW in their premium class cars in 2001[35]. The "iDrive" (and similar control concepts of other automobile manufacturers, e. g. Renault's "joystick" [253, p. 15]) shows that the usage of the haptic modality has potential for offloading screen communication and increasing perceptual bandwidth available for interaction with a mobile information appliance.

(*ii*) BMW's Lane Departure Warning (LDW) system alerts the driver to potential lane departure by providing haptic warnings through vibration in the steering wheel[36].

This approach is similar to the raised pavement markers (particularly in the US known as "Butt's dots") or rumble strips, a road safety feature separating the lanes of a road and causing tactile feedback (vibrations) on the steering wheel when driven over.

(*iii*) Automobile manufacturer Citroen has been integrating a Lane Departure Warning system based on vibro-tactile notifications via the car seat since 2004 [254], [255, p. 34].

(*iv*) Audi's recent collision warning system provided in the premium class informs the driver in case of collison risks with a warning jolt produced by the brake system[37].

[35]The control device was first demonstrated at the International Motor Show (IAA) in Frankfurt, Germany, September 16 to 26, 1999.

[36]Siemens VDO and Mobileye make LDW available for BMW's 5-series, published online March 2, 2007, http://mobileye.com/sites/mobileye.com/files/SVDO.ME.LDW.pdf, last retrieved August 30, 2009.

[37]Audi Q7, http://www.audiworld.com/news/05/frankfurt/q7/content4.shtml, last retrieved August 30, 2009.

(*v*) "Touchy-Feely Screen" (Immersion Corporation)[38] is a touch-screen with haptic feedback provided by precise motors vibrating in the top layer of the display. The vibration varies depending on the region (or graphic) touched on the screen (application in vehicles e. g. for the air-conditioning control or multimedia system).

(*vi*) A 2004 report from Denso Corporation, Japan, indicates that the majority of commercial vehicles will be equipped with some vibro-tactile stimulation device by 2020 [220, p. 656].

(*vii*) Forward Collision Warning (FCW) systems are available on trucks and buses running in the USA. They are useful when the environmental conditions are critical (for example when it is foggy or when the driver's judgement is limited or even wrong), because they are able to detect objects ahead of the car, and in case there are dangerous obstacles to warn the driver so that corrective actions can be taken [35, p. 771].

6.7.3 The Driver Seat: Suitable for Haptic Notifications?

A vehicle seat, particularly that of the driver, belongs to the most important components of a car, not only with regard to comfort – professional drivers spend most of their time there[39] –, but moreover in improving the work environment of the car driver, and thus have been the subject of intense research interest in recent years [251]. Furthermore, the vehicle seat accommodates large regions of the driver's body and is always in contact with the driving person, and therefore is best suited for being used as a medium of interaction (data carrier) for both driver-vehicle input and vehicle-driver output.

However, it also has to be considered that the seat already relays most of the vibrations generated when a car is operated towards the driver [256]. The European Union Directive 2002/44/EC (also referred to as the "Vibration Directive") places responsibilities on employers to ensure that risks from whole-body vibration are eliminated or reduced to a minimum. Whole-body vibration in this context is caused by vibrations transmitted through the vehicle seat to the driver or passenger. Exposure to high levels of whole-body vibration can present risks to health and safety. The effects are worst when the vibration intensity is high, the exposure durations

[38] Technology Review, published online September 2005, http://www.technologyreview.com/Infotech/14751/?a=f, last retrieved August 30, 2009.

[39] According to European Union Regulation (EC) No. 561/2006, http://ec.europa.eu/transport/road/policy/social_provision/social_driving_time_en.htm, last retrieved August 30, 2009, the daily driving time shall normally not exceed 9 hours, although the daily driving limit may be extended to at most 10 hours (not more than twice during the week). A weekly driving time limit of 56 hours is the maximum permitted. But as the past has shown, these limits are often exceeded and have led to serious accidents caused by overfatigue of the driver.

long and frequent, and the vibration includes severe jolts [257, p. 6]. However, Schust *et al.* [258, p. 622] investigated the influence of vibrations on the performance of vehicle steering tasks and found – contrary to their expectations – that there is no significant effect of increasing vibration intensity, compared to reaction times in motor tasks such as braking or accelerating.

Niekerk *et al.* [259] investigated the sources of vibration, too. They distinguished between road input via tire contact and vibrations induced from the power train. Wu *et al.* [260, p. 939] confirmed these two sources of vibration in their research work on Active Vibration Control (AVC) for reducing undesired small-amplitude vertical vibration in the driver's seat of a vehicle. An extensive study of the progress of the parameters fatigue, discomfort, and performance of drivers while sitting on a (vibrating) driver seat for a long time has been presented by Falou *et al.* [261]. They found that (*i*) subjects became increasingly uncomfortable during the trial of 150 minutes (true for all experimental conditions), and (*ii*) driver performance was worst when sitting in an uncomfortable seat with presence of vibrations, and best when sitting in a comfortable seat with absence of vibrations. Other studies focusing on vibration distribution in vehicles and their impacts on driver discomfort (namely fatigue and safety) have been presented for instance by Paddan *et al.* [262], Jang and Griffin [263], and Demic *et al.* [264].

General Conclusions

The exploration of related work and the research conducted in the scope of this thesis showed that the application of the sense of *touch* for supporting the *visual* and *auditory* sensory channels appears to be a promising step towards more convenient, relaxed, and safe future driving behavior.

Using the *haptic* sensory channel allows feedback to be transmitted immediately to the driving person – with the benefit of affecting the driver's cognitive workload to a lesser degree than when using *visual* or *auditory* feedback [265]. Touch-based expression of drivers' behaviour has the potential for novel, implicit input. The first public document released by the European CAR2CAR Communication Consortium in August 2007 proposed using the *haptic* channel, beside the *visual* and *auditory* ones, for expression and feedback of almost any kind of application (e. g. for collision or traffic jam warnings, slow vehicle ahead warnings, traffic optimization and routing systems, access control, remote diagnostics, personal data synchronization, etc.) [266].

In order to successfully deliver vibro-tactile information to the seat and the driver, and consequently to ensure reliability, stability, and accuracy of that touch-based system, it is absolutely essential to take the (whole-body) vibrations into account, as vibro-tactile information normally is superimposed upon these basic car-induced vibrations. A feasible approach for reducing the negative effects of whole-body vibrations has been presented by Frechin *et al.* [267] – their system called "ACTISEAT" is an active seat, isolating passengers as well as in-vehicle equipment from vibrations and furthermore compensating for acceleration in all directions to a certain extent.

Aside from sensor mats and vibro-tactile actuators integrated into the car seat – as proposed in this research work – much potential can also be seen in using other control devices a driver is interacting with, such as for instance the steering wheel (where the hands and the fingers of the driver effect a great part of the control of the car [268]), the gear-shift lever, and the emergency brake.

"I am among those who think that science has great beauty. A scientist in his laboratory is not only a technician: he is also a child placed before natural phenomena which impress him like a fairy tale."

Marie Curie (1867 – 1934)

French chemist and physicist

Part III

Information Needs of Drivers

7 The Driver as the Weak Point in Interaction

The car driver is believed to be the most prominent factor in concerns regarding the safety of vehicle control or road traffic – according to Treat *et al.* [269], approximately 70 – 90% of all accidents are the fault of the driver to a certain degree. Apart from factors like motor and cognitive abilities, fatigue, attitude, or alcohol use [270], the interaction between a driver and the vehicle itself has increasingly become the weak point of safety. The automobile (sub)systems are getting more and more reliable, safe, and comfortable due to technological advancement and experience in system and user interface design.

The driver is challenged to adapt rapidly to new control concepts, as it is well known that new DAS and other information systems appear overnight and Human-Computer Interfaces are revised constantly. Furthermore, the driving person must deal with increasingly complex systems. Any combination of these factors may result in an excessive cognitive load for the driver, and subsequently to the accepted fact that the driving person is the weak point in the Driver-Vehicle Interaction chain.

As a consequence, an increasing effort is required to improve in-time and accurate driver-to-vehicle communication with as low as possible a cognitive burden for the driver. This is a great challenge which was not considered by automobile manufacturers in the past.

This part is dedicated to questions like (*i*) what kind of information is essential for a driver to guarantee safe vehicle operation?, (*ii*) at which time, or in which segment of time, should the information be provided?, (*iii*) what is the minimum information update interval?, (*iv*) what is the best modality to notify the driver?, or (*v*) where should the information be presented?, with the aim of providing suitable solutions or recommendations for improving driver interfaces.

7.1 Cognitive Load: A Big Challenge

The ever increasing emergence of information sources in and around the automobile necessitates, alongside the main activity of vehicle control, an interactive effort from the driver.

The level of attention required is typically expressed as *cognitive load*. It can be described as the demand placed on a user's working memory during a task or "the amount of mental resources needed to perform a given task [..]" [271, p. 421].

There are many situations where the demand on a person exceeds its capabilities, and thus results in performance deterioration [272, p. 1]. Examples are that (*i*) users have to deal with too much information, (*ii*) a larger number of tasks have to be completed in a given time, or (*iii*) the time within which tasks can be performed is reduced [272]. To get an estimation about the task complexity, an objective analysis of the tasks themselves is required (and is actually used in cognitive load theory) [273, p. 61].

According to Jonsson [271], the definition of cognitive load is the amount of mental resources needed to perform a given task (taken from Paas and Van Merrienboer [274], [275], and Sweller *et al.* [276]). More detailed, cognitive load is estimated by the number of statements and productions necessary to be handled in memory during a specific task; drivers, for instance, use cognitive resources for attention, memory and decision making while driving and complex traffic situations and unfamiliar road designs require additional resources [271, p. 421].

This calculation gives a quantitative estimate of task difficulty. In the domain of Human-Computer Interaction, Baber *et al.* [272] have focused on the behavior resulting in temporary overloads – for this, the detection of overloads is needed first, e. g. with Bayesian network methods [273].

Knowledge of cognitive abilities and actual workload is important for the driving task, since even the smallest distraction or diversion of cognitive resources can have disastrous effects [271, p. 421].

7.2 Empirical Evidence for Cognitive Load

A driver distracted from the main activity of driving[40] due to on-board services like infotainment or interactive equipment (and there is evidence that these systems increase driver distraction and driver workload, [65, p. 256]), perhaps reacts belatedly to (important) requests, or fails to register them at all [277, p. 21].

Research on car following indicates that when the lead car suddenly decelerates, drivers performing a cognitive distraction task (such as cell phone conversations) take longer to release the accelerator pedal, apply the brake and decelerate (Lamble *et al.* [278], Hurwitz *et al.* [279], Lee *et al.* [280]). Wood and Hurwitz [281] experimented with cell phone conversations during demanding driving situations, and tested if intelligently suspending the conversations would improve the driver's performance and lessen his/her subjective workload. Their results show that the conversation intensity has a significant effect on driver workload. They found that driving was mentally and temporally demanding, required more effort and was more frustrating, and finally that the driving performance suffered while engaged in an intense conversation.

Another study, presented by Cooper *et al.* [282], reports that distracted drivers accepted shorter gaps when making left turns. Hanbluk *et al.* [277] experimented on the performance of demanding cognitive activities (such as Driver-Vehicle Interactions with speech control, or in using hands-free in-car devices). They found that the performance of a demanding cognitive task while driving produced changes in (*i*) drivers' visual behavior, (*ii*) vehicle control, and (*iii*) subjective assessments of workload, safety, and distraction; with increased cognitive demand, over half of the drivers changed their inspection patterns of the forward view.

[40]The task of driving is often equated with keeping one's attention on the road and the surrounding environment.

According to Hoedemaeker *et al.* [165], the three parameters (*i*) time pressure, (*ii*) level of information processing, and (*iii*) task switching define the critical areas of cognitive load in vehicles[41]. They suggest using adaptive interfaces, which have the capability to estimate the momentary load on the driver via sensing state and behavior of the driver, resources of information, and environmental conditions, and adapt service and information provisions in order to change the load on these three dimensions and consequently on the driver (for instance, filter information to minimize task switching). Furthermore, they ensure that the information from the different services is presented to the driver without causing cognitive overload or negative consequences for traffic safety.

7.3 Managing Workload

Driver-Vehicle Interaction management is currently a hot topic in the automotive industry [284]. A structured way to deal with Human-Machine Interaction (HMI) issues arising from the rapid functional growth in today's vehicles (such as overload, distraction, steering failures, accidents) is the application of a new type of "assistance system": The *Workload Manager* (also known as *Interaction Manager*, or *Adaptive Interfaces*) with two key driving forces (*i*) the need to handle the rapid in-vehicle functional growth[42], and (*ii*) expected benefits in terms of road safety [284, p. 6].

In detail, a workload manager is a system that attempts to determine if a driver is overloaded or distracted by continuously assessing the difficulty or complexity of driving and regulating the flow of information to drivers that could interfere with driving (filtering and prioritizing the information), such as incoming phone calls [285]. The (potential) difficulty of a certain driving situation is assessed by vehicle sensors (such as speed, braking, headlight and windshield wiper usage).

The economic actuality is endorsed by Engstroms and Flobergs 2008 European Patent EP1512584 [286] for instance – they just recently protected a "method and device for estimating the workload for a driver of a vehicle". The patent describes a technique to indicate the driver's attention and the state of the vehicle and determine the workload level of the driver with a workload estimator. This device further includes means for inhibiting access to incoming calls, text messages and/or internally generated messages for a period of time where the workload is high.

[41] These parameters are based on the Cognitive Task Load (CTL) theory, first developed by Neerincx [283].

[42] Not only the integration of sensors and computing hardware, but also the Driver-Vehicle Interaction side needs to be addressed, e. g. strategies for message prioritization, interface consistency, etc.

7.3.1 Examples

Since 2003, most Saab 9-3 and 9-5 models in the United States are equipped with a workload manager called "Dialog Manager", which suppresses certain information displays during demanding driving conditions. A similar system, called the "Intelligent Driver Information System (IDIS)", is available in Volvo's S40s and V40s sold in Europe since 2004 [284]. IDIS blocks telephone calls and text messages during times when the driver is turning, changing lanes, or conducting similar maneuvers [287].

8 Driver Activity and Notification Demands

The number (particularly that of non-driving tasks) and the complexity of activities inside the car – as shown in Fig. 8.1 – has risen constantly in the last years. One possible consequence of this development is an increase in driver distraction [99] which needs to be avoided or compensated as much as possible in order to ensure safe and accident-free vehicle control. A second issue is that most of the vehicular systems (see righthand side of Fig. 8.1) deliver feedback on the car's status (which cannot be switched off) to the driving person and thus, lead to information overload and driver distraction.

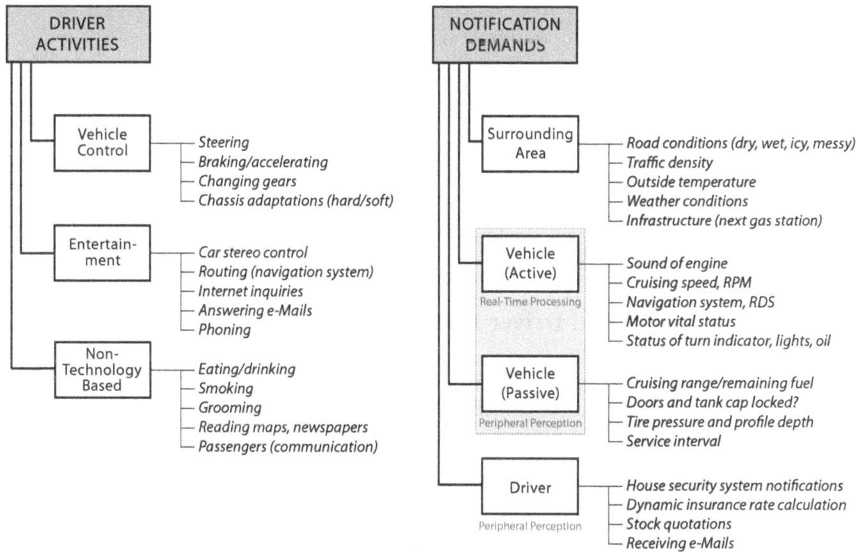

Fig. 8.1: Driver activities and notification demands when operating a car.

Since different activities require more or less active cooperation of the driver (and therefore result in a higher or lower level of distraction), and occur infrequently (internet inquiries) or often (navigation), a universal, appropriate and comfortable solution has to consider the required level of attention for a certain notification demand (such as ignore, interest, action) while taking the driver's current mental state (e. g. smooth, stressed, heavily disturbed, etc.) into account – the concept of raising a certain attention-level is introduced in the Section "Level of Attention (LOA)" on p. 101.

Further potential for decreasing driver distraction is attributed to multimodal, instead of unimodal, feedback. A detailed discussion on multimodality is given in both Chapter "Multimodal" on p. 21 and in Vernier *et al.* [288].

Allen [5] has provided a means of combining multiple modalities into a single composite modality as shown in Fig. 8.2. To do so, he addressed the problem of reasoning about qualitative temporal information and in particular gave an algorithm for computing an approximation to the strongest implied relationship for each pair of intervals[43].

Combination Aspects	Combination Schemas				
TEMPORAL	Anachronism	Sequence	Concomitance	Coincidence	Parallelism
SPATIAL	Separation	Adjacency	Intersection	Overlaid	Collocation
SYNTACTIC	Difference	Completion	Divergence	Extension	Twin
SEMANTIC	Concurrency	Complementary	Complementary and Redundancy	Partial Redundancy	Total Redundancy

Fig. 8.2: The five *Allen relationships* applied to the four combination aspects temporal, spatial, syntactic, and semantic (adapted from Allen [5]).

8.1 Notification-Induced Driver Distraction

The driving person is continually burdened with notifications from the different vehicular systems, the environment and requests initiated by the driver himself. The vast amount of information delivered to the driver is probably a reason for distraction and therefore attention must be focused on this topic in order to prevent or compensate for distraction.

Distracted driving is part of the broader category of driver inattention[44] and occurs "when a driver is delayed in the recognition of information needed to safely accomplish the driving task because some event, activity, object or person [..] compelled or tended to induce the driver's shifting attention away from the driving task" (American Automobile Association Foundation for Traffic Safety) [289, p. 21], [290], [270].

Royal [291, p. 3] states that about 3.5% of all drivers in the US (up to 8.3 million) that have been involved in a crash attribute it to them being distraced. The National Highway Traffic Safety Administration (NHTSA) estimated that driver distraction contributes to approximately 13% of crashes reported to the police (Young *et al.*, [290, p. V]). In 2004, automobile manufacturer General Motors issued a public statement which suggested that driver distraction

[43] The five relationships presented here, together with the four inverse relationships, are a subset of all thirteen possible relationships in which an ordered pair of intervals can be related. Originally the temporal relation "coincidence" was segmented into three relations [5, p. 4].

[44] Inattention refers to any condition, state or event that causes a driver to pay less attention than required for the driving task [270].

contributes to 25% of crashes; furthermore, they stated that the most important principle for avoiding these accidents has to involve minimizing "eyes-off-road time" and "hands-off-wheel time" – the two basic automotive principles. Estimates from other sources are as high as 35 – 50% (Stutts *et al.*, [292]). However, as more wireless communication, entertainment and Driver Assistance Systems proliferate on the vehicle market, it is probable that the rate of distraction-related crashes will increase (Stutts *et al.*, [292]). Nevertheless, almost no research has been conducted in the field of (automatically) detecting the driver's inattenation [293]. One method, presented by Torkkola *et al.* [294], used primary task parameters, such as the steering wheel angle or the position of accelerator or brake pedal to detect (and compensate for) inattentiveness.

8.1.1 Causes of Distraction

Distraction of vehicle drivers is either caused by the driver him/herself (fatigue, drunkenness) or results from the vehicle (type, condition), or from environmental factors (weather, road conditions, traffic situation, etc.). Of particular interest are distractions deriving from the driver's behavior; they are produced by (*i*) something a driver sees or hears (or potentially, when employing the sense of touch, feels), (*ii*) some secondary driving task (such as eating, drinking, operating the car stereo, etc.), or (*iii*) social activities (conversations with passengers, cell phone calls). A more detailed explanation of the different classes of distraction is given, for instance, in the work by Stutts *et al.* [295, p. III-1] or in [67, p. 29].

According to the NHTSA, four distinct, but not mutually exclusive, types of distractions can be distinguished: (*i*) visual, (*ii*) auditory, (*iii*) physical, and (*iv*) cognitive distractions [290]. The cell phone would be one example, incorporating all four distraction-forms, namely looking at the phone to dial a number (visual), holding a conversation with someone (auditory), dialing a phone number (physical), and focusing on the conversation rather than on steering the car and observing the environment (cognitive).

It would be practical to further diversify the sources of distraction, e. g. into driver-influenced, vehicle-centered and environmentally-caused distractions, or into technology-based and non-technology-based distractions. One example of an environmentally-caused distraction (arising from the driver's physical limitations) is driving through a road tunnel. Particularly the sensory modality *vision* is affected by environmental factors, such as glaring or reflecting light and by the different abilities of day and night vision. According to the adaptability of the eyes, and to avoid distractions or accidents, the luminance density between tunnel entry and exit varies (Fig. 8.3). This variation of lighting or luminance density in road tunnels is standardized, e. g. for Germany in DIN[45] 67524 "Lighting of road tunnels and underpasses".

[45]DIN=Deutsches Institut für Normung (German Institute for Standardization).

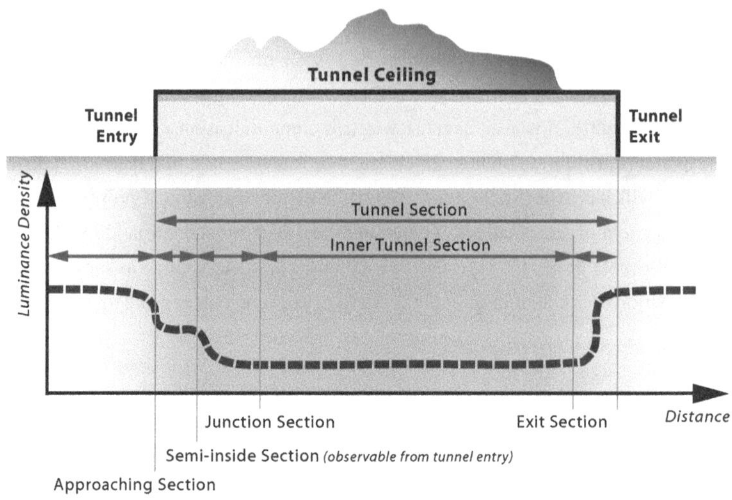

Fig. 8.3: Luminance density in road tunnels.

8.2 What Information?

Information brokerage in vehicles, and particularly managing the information flow from the vehicle to the driver is becoming increasingly important due to the large amount of information emerging from vehicular assistance systems (an incomplete categorized list of possible information is shown in Fig. 8.1 on p. 71). There is a strong prerequisite that the next generations of these (interactive) systems operate without decreasing the safety or ease of operation of vehicles and increasing drivers' workload, respectively. In order to guarantee this, integration within a human operation context is required to give priority to the needs of the driver, rather than to the characteristics of hardware, software and infrastructure [6].

The importance of intuitive, non-distracting Driver-Vehicle Interfaces as well as the task-centered presentation of information has also been picked up by Richard Parry-Jones (Ford Motor Company, Product Development) [296] – among others, he suggested the following points:

(*i*) Reliability is the greatest challenge to infotronics[46] development, but the man-machine interface is a close second.

[46]Infotronics is the area of hardware-oriented informatics and embedded systems design. In addition, it covers the development of intelligent sensors and actuators, as well as signal processing on distributed systems, miniaturization and resource optimizations. Source: Lucern University of Applied Sciences and Arts, Technics and Architecture (HTA), Department of Electronics, Technikumstrasse 21, CH-6048 Horw, http://english.hslu.ch/, last retrieved August 31, 2009.

Potential Intelligent Transportation Systems (ITS) In-vehicle Information		
SAFETY and CAS (Collision Avoidance Systems)	ATIS (Advanced Traveler Information Systems)	CONVENIENT and ENTERTAINMENT SYSTEMS
Road Departure	Trip Planning	Telefax
Rear End	Route Guidance	Pager
Lane Change/Merge	Route Selection	Radio-/CD-based Audio Systems
Intersection	Multimodal Coordination	Cell Phone
Railroad Crossing	Route Navigation	Television (DVD)
Drowsy Driver	Yellow Pages	Mobile PC (PDA)
Automatic Cruise Control	Automated Tolls	Retrievable Settings for Seat and Mirrors
Yaw Control	Motorist Services	Personal Messages (E-Mail)
Roadside and Emergency Services	Vehicle Status	Stock Quotations
Vehicle Location and Voice Availability	Regulatory Information	House Security System Status
Stolen Vehicle Location System	Travel Advisories	
Theft Detection	Road Condition	
Road Sign Detection (e.g. Opel Eye)	GPS	
Blindspot Vehicle Detection	Personal Car-Insurance	
Overtaking Warning System	Caused-based Taxation	
	Geo-Information Services	

Fig. 8.4: Categorization of driver information in vehicles into three classes (adapted from Kantowitz and Moyer [6, p. 3]).

(*ii*) Infotronic technology must be intuitive and easy-to-learn.

(*iii*) Next-generation interactive systems must blend information, communication, and entertainment technologies into the everyday driving experience without complicating the basics of operating vehicles.

8.2.1 Classes of Information

According to Kantowitz *et al.* [6] there is a strong need to integrate the three classes (*i*) safety systems and CAS, (*ii*) Advanced Driver Assistance Systems (ADAS), and (*iii*) convenience and entertainment systems into the vehicle. Fig. 8.4 gives an incomplete but representative list of appliances or systems for each of the three in-vehicle information classes, currently available in vehicles or shortly before market introduction. It should be noted that the aggregation of information available from these systems would exceed a driver's information perception and processing capacity by far.

A perfect architecture for information provision in vehicles from the driver-centered perspective will place no limitations on the driver's ability to obtain desired information, and simultaneously avoid stress, cognitive overload, or distractions from the primary task of driving by preventing the emergence of too much information. Using only the two traditional sensory modalities *vision* and *speech* for information transmission would not solve the problem of distraction and cognitive overload. It is hoped that incorporating additional senses, and in particular the sense of *touch*, will help to reduce drivers' cognitive load.

8.3 When and How?

The question *when* in the meaning of "the perfect point in time" cannot be answered universally for all kinds of arising notification demands. Moreover, *when* depends not only on the information of a specific system itself, but also on a number of factors, first of all the kind (or the importance) of information[47].

For each in-vehicle system having the capability of delivering information to the driver a notification factor in the range [0..1] should be determined (where 0.0 means that the information is not important and can be ignored by the user, or the information display can be suppressed, and 1.0 means that the user should be absolutely aware of the information and that the driver's attention is fully occupied due to the criticality of the information).

A second issue is the modality or the modalities used for transmitting the information to the driver. Each modality (*vision*, *hearing*, *touch*, etc.) has its own temporal behavior, and the usage of multimodal information transfer instead of the unimodal one necessitates separate considerations too. Different types of information are better suited for being presented with the one or the other modality (general guidelines for when to use a designated modality, as defined by Simpson *et al.* or Deatherage, are given in [166, p. 30]).

A third factor results from the number of concurrently presented information items. It is important to note that the notification level of each information item plays an important role (only one or few items with notification level 1.0 can be presented at one time, while several notifications with a lower level can be transmitted).

A further factor, assessing the dynamics of information presentation, is that some information requires higher update rates (e. g. cruising speed, notification about persons approaching on any side of the car, driving in the wrong lane, a blind spot warning system, etc.), and others are less

[47]In accordance with Matthews *et al.* [297, p. 249] the term *notification level* is subsequently used instead of importance of information. Higher notification levels correspond to more critical data and demand a user's full attention. Lower notification levels correspond to non-critical data and are typically displayed in a way that does not attract attention, but allows a user to monitor the information display peripherally.

important and thus, a lower update rate or even a peripheral perception would be sufficient (for instance, coolant temperature, charge-level of the battery, fuel status, outside temperature).

8.3.1 Information Capacity Limits

The maximum capacity of information perceivable for the driver in a certain moment may depend on several, even person dependent, parameters as well as on the type and number of sensory channels utilized and necessitates the definition of a personal boundary (e. g. an experimental determination) which must not be exceeded in order to ensure safe vehicle operation.

For example, for the eye, Laurutis [298] investigated the limit of the amount of information transmitted through the visual channel by adapting the information theory (for details see Shannon [299]) to this bioengineering field of application. Theoretical and experimental analysis of the information capacity of the human's visual system (for smooth pursuit and saccadic eye movements) showed that this parameter is useful and could also be applied to define quantity parameters for other Human-Machine Interaction channels.

According to these results, the information capacity limit for a specific driver has to be dynamically adjusted as a function of the actual driving situation and the driver's mental state.

8.3.2 Selection of Sensory Modalities

Correctly choosing the sensory modality (*hearing*, *vision*, or even *touch*) for different display demands in vehicles is important in order to guarantee optimal performance or effectiveness of information presentation. One important assertion is that the visual attention demand must be minimized to enhance the safety of these information systems [166, p. 10]. As an alternative, particularly in situations with high visual attention demand, the auditory or tactile modality should be utilized. As information in the driving environment is often direction-based (e. g. danger from the left), the omni-directional nature of auditory and haptic displays makes them most desirable for alert and warning messages.

Visual Notifications

As an example, the parameters for visual information presentation are discussed in more detail. French [300] and Labiale [301] investigated safety implications associated with driver information systems. The major concern of their research was the average glance time that is considered safe when the driver is looking at an in-vehicle display. They found that a driver's average glance time is $1.28s$. Additionally, French [300] indicated that glance times greater than $2.0s$ are unsafe and unacceptable, glance times between $1.0s$ and $2.0s$ are considered marginally safe, and glance times under $1.0s$ are considered safe. Furthermore, Labiale [301] found that 92.3% of all glance times were less than $2.0s$.

Auditory Messages

Auditory message length refers to the time required to transmit a number of words or sentences necessary for presenting auditory information to the driver. Depending on the type and importance of the presented information, the message length can be varied. There is evidence that the longer the message is, the more processing time is required by the driver. As a consequence, messages that require the driver to respond immediately should be as short as possible. One-word messages informing the driver of the appropriate action to take might work best in situations such as these. Particularly for longer messages, an effort should still be made to make the messages as concise as possible.

According to Campbell *et al.* [7] the following empirically validated design guidelines for information displays are recommended:

(*i*) Messages that require an urgent action should be a single word or a short sentence with the fewest number of syllables possible. Drivers should be able to understand the message immediately.

(*ii*) Messages that are not urgent or for which a response may be delayed can be a maximum of 7 units of information in the fewest number of words possible. If the information cannot be presented in a short sentence, the most important information should be presented at the beginning and/or the end of the message.

(*iii*) Navigation instructions should be limited to 3 or 4 information units (for example, the topmost suggested message in Fig. 8.5 consists of 3 information units: Accident ahead, merge right).

Examples of Auditory Messages	
SUGGESTED	NOT SUGGESTED
"Accident ahead, merge right."	"There is an accident ahead in the left lane. Merge right as soon as possible."
"Turn left."	"Esteemed driver, at the next crossing please prepare to turn left."
"Oil change needed by November 15, 2009."	"The vehicle maintenance log shows, that the oil change for your vehicle is due. Please complete the change by November 15, 2009."

Fig. 8.5: Examples of suggested and not suggested auditory message length (adapted from Campbell *et al.* [7]).

Sense of Touch

A detailed analysis of the different types of senses (as shown in Fig. C.1 on p. 214) and their individual usability shows that haptics offers great potential as a cognitive load reducing interaction modality, but was given only little consideration in the corresponding research area until now. As a fact, there are no commercial products available in the automotive domain which consider the full range of haptic interaction.

The application of haptics in order to signal an occurrence is able to relieve the cognitive load on *visual* (eyes) and *auditory* (ears) channels of information. However, previous experiments (discussed in detail in the Section "Experiments" on p. 119) have shown that intuitive and unmistakable identification of many haptic patterns is not easy or even impossible. Furthermore, additional disorientation would arise if there is no standardized mapping for a specific informational item and the corresponding haptic sign or symbol – as is common in the market economy, each (seat) manufacturer would define its own standard. In order to ensure accurate, reliable and universal usage of haptic feedback, the definition of a easily learnable *vibro-tactile language* or a *haptic alphabet*, which could be similar in its definition to the sign language or the Braille alphabet, would be required

As a result of advancements in tactor hardware and in control logic, it should now be possible to define an intuitive haptic alphabet, probably with a larger character repertoire than the "Vibratese language", developed by Geldard *et al.* [23] in the 1950s.

8.3.3 Airplanes – The Better Cars?

As information items and control switches inside the vehicle grow, functions that enable drivers to access information that was previously unavailable, or was available only in domains outside the vehicle, become available. When giving drivers access to these ADAS inside the vehicle, designers must consider certain points, such as safety and usability – otherwise it would be possible that these helping systems overload the driver and reduce any safety benefits that could be gained from their use [166]. With increasing complexity and a larger number of in-vehicle systems, dashboard and interface designers have started to copy some of the interface paradigms of airplanes. In the aviation domain, many lessons have been learned that can be applied to the integration of in-vehicle information, including operator workload research [6, p. 1].

Multimodality

Driver-Vehicle Interfaces could be improved by incorporating multimodal stimuli (a combination of *auditory*, *visual*, and *tactile* stimuli). Dingus *et al.* [302] recommended (*i*) using the auditory channel for providing an auditory prompt to look at a visual display of changing or upcoming information (thus lessening the need for the driver to scan the visual display constantly in preparation for an upcoming event), and (*ii*) having some type of simple visual information presentation to supplement the auditory message (so that a message that is not fully understood or remembered can either be checked or referred to later via the visual display).

8.4 Where?

Each modality of information presentation in vehicles requires its own considerations about where to present the information so that the driving person can perceive it optimally.

As the driver is looking at the road most of the time while driving, a *visual* display should be placed near the windshield (or better in the windshield; a technique known as *head-up display*, projecting status information into the windshield, has been commercially available, e. g. in BMW's premium class, since 2003 [303]). The duration of the distraction from the main task is affected by two factors. On the one hand, the view is actually turned away from the road and is directed toward the announcements in the instrument panel, which takes some time, and on the other hand the accommodation of the eye – changing the focus from far (the road) to near (the dashboard) – takes time. Particularly older drivers are strongly affected by this focusing task, which results in symptoms of fatigue.

Auditory displays are omni-directional and information from that kind of display can be obtained everywhere in the car (this could also be a disadvantage, as passengers in the rear seats would be informed about things not intended for them). For auditory displays it is important to consider other sources of sound, such as motor noise, environmental sound, or the voices of passengers, and correspondingly adapt the transmission of auditory information (avoid or compensate overlays that make it impossible for the driver to interpret system notifications, especially if they are mandatory).

For *haptic* notifications, a number of perceivable, feelable, or graspable devices in the car exist – all of them allowing the integration of vibro-tactile actuators for vehicle-driver information presentation, for instance the driver seat and/or back, the steering wheel, the hand brake, throttle or brake pedals or the clutch. For broader or universal usage, the seat would be best suited, because (*i*) it is substantial large-sized, thus allowing notification via large parts of the body, (*ii*) it allows, again due to its size, the integration of hundreds or thousands of sensors and actuators, (*iii*) it has sufficient space to accommodate the control computer hardware, (*iv*) it can easily be connected to car buses (FlexRay, CAN), which provide for secure operation and supply access to other sensors in the vehicle, and (*v*) it is attached and available in any car (or other type of vehicle).

8.5 Visual and Auditory Perception

In this paragraph, a summary about the significance of the different sensory modalities and their potential for driver distraction, cognitive overload, or casualties for in-car usage is given. The conclusion of the short study should be used in the next parts of this work for the evaluation of novel approaches for the future of ADAS toward a more convenient, stress-free, and safe driving behavior.

Vision is the modality most frequently used for information presentation in vehicles [75, p. 130], [25], followed by *auditory* information [50, p. 41]. As driving is a visual task requiring the driver's eyes as well as his/her visual processing capacity on the road most of the time, the remaining *visual* capacity should not be used exclusively for obtaining important information from control instruments or the numerous Driver Assistance Systems [304, p. 42]. Lansdown [305] has substantiated that a congestion warning device (which demands attention continuously) represents a significantly greater visual demand and a subjectively higher workload than an in-car entertainment system or other unimportant control instruments (constant monitoring of these displays is generally not required) when observed in a road trial. In [306, p. 4], Green *et al.* stated that information presentation from a collision avoidance or obstacle detection system on a visual display may generate additional driver distraction at a time when vision (on the road) would be of particular importance.

Auditory information presentation can be used instead of *visual* displays in order to relieve drivers' workload [307]. Liu [16] evaluated the performance of navigation and button-push tasks using visual, auditory, and multimodal (visual and auditory) displays. The experiments showed that both the auditory and multimodal displays produced better response times than those using the visual-only display. Furthermore, the visual display led to less safe driving (caused by a higher demand on the driver's attention).

In a study of related work, Liu [16] presented some more substantiated statements relevant for this work:

(*i*) Auditory navigation devices of any complexity make significantly fewer navigation related errors than visual devices.

(*ii*) Drivers using an auditory navigation device reduced travel distance and travel time.

(*iii*) Using auditory displays for navigation information reduces drivers' workload compared to systems using visual displays.

(*iv*) Auditory information presentation is preferred by drivers.

(*v*) Drivers using an auditory display in high workload driving situations did not reduce their speed as much as those using visual devices.

(*vi*) An auditory display will improve time-sharing performance in heavily loaded visual display environments (according to Wicken's MRT, [59]).

(*vii*) Multimodal displays allow drivers to perceive more information without significantly increasing their workload.

8.5.1 Summary

According to Wicken's Multiple Resource Theory (MRT) [59], the different sensory modalities used in in-vehicle interaction interfere with the car driver to a certain degree.

When having a closer look at the results from the list above it turned out that using the *auditory* modality has the potential to reduce drivers' cognitive load and to increase road traffic safety. Seppelt *et al.* [308, p. 6] quotes a study by Folds and Gerth which indicates that *auditory* presentation of warning information led to faster and more accurate primary and secondary task performance than *visual* presentation of the same information. Szalma *et al.* [309] reported that the speed and accuracy of signal detections are greater, and there is less deterioration of the vigilance level with *auditory* than with *visual* signals. Vollrath and Totzke [310, p. 6] found that it is preferable to present information in vehicles with the *auditory* channel and avoid information presentation with the *visual* modality, because it impairs the driving performance significantly. The general conclusions drawn from studies regarding *visual* and *auditory* task performance are that performance with the *visual* modality suffers from increased spatial separation from the primary driving task – which gives the *auditory* modality the advantage of functioning better than the *visual* modality [308].

However, on the other hand, it has also been verified that *auditory* information may interfere with the task of vehicle steering. In an experimental study, Wheatley *et al.* [163] found that 53% of the subjects felt that an auditory warning tone actually interfered with their ability to drive safely and 59% indicated that it was more difficult to concentrate on driving. Wiese *et al.* [311] showed that drivers' annoyance associated with highly urgent sounds increased their workload (a strong positive association between ratings of annoyance and subjective workload).

8.5.2 Research Potential

The investigation into using the *tactile* sensory channel in vehicles for both unimodal and multimodal interaction offers great potential, but necessitates consideration of several questions. A theoretical and experimental assessment of these open issues is the main contribution of this work and will be presented in the next sections.

9 Advanced Driver Assistance Systems (ADAS)

This chapter is focused on a detailed examination of the sense of *touch* (or *haptics*) as a promising technology for enhancing Driver Assistance Systems, starting with an overview of the state of the art and emerging Advanced Driver Assistance Systems. Other research prospects, such as context-aware services [312] or the utilization of the two sensory modalities *smell* and *taste* are left unaccounted in this work.

Driving a vehicle has changed relatively little since the first cars appeared on the road, but with rapid advances in In-Vehicle Information Systems and Advanced Driver Assistance Systems this task will now be transformed (see Fig. 9.2 for an overview ADAS available today). One of the major reasons for the introduction of these driver-support systems is to cope with the various vehicle handling risks, originating from the larger number of cars on the streets, the complex design and layout of roads, a stressful life style, etc. (see Fig. 9.1 for a rough classification of driver risks).

Classes of Driver Risk		
TRIP PRECURSORS	EXTERNAL EVENTS	DRIVER BEHAVIOR
- Drinking - Fatigue - Vehicle maintenance - Shortage of time - Etc.	- Road geometry - Traffic jam/density - Road accidents - Weather conditions	- Speed choice - Gap acceptance - Headway - Eyes off the road

Fig. 9.1: Aspects of driver risks (adapted from Lerner *et al.* [8]).

Thus, ADAS have been designed to lead to a safer and stress-free driving experience (Philips [313], Hella KG [314, p. 9], Nirschl [315], Goroncy [316]). Because most of the ADAS listed in Fig. 9.2 are state of the art or well-known, replicating information regarding their functionality is omitted here (a good starting point for detailed information would be Bishop *et al.* [9] or Heijden and Marchau [10]).

The expected potential of Advanced Driver Assistance Systems to improve vehicle-handling performance, relieve driving persons of distraction or reduce their workload, lower traffic hazards, reduce or even eliminate the errors of the driver and enhance driving efficiency [317, p. 247] increasingly attracts the attention of car manufacturers and important decision makers to integrate new devices, or refine existing systems. For example, using an in-vehicle navigation systems reduces the driver's attention to the task of navigating and searching for alternative routes, or automatic cruise control maintain the car's speed and decrease the driver's load [318].

However, there are also potential problems to be expected. With an increasing complexity of in-vehicle tasks and display systems, information overload has become a growing problem for drivers. Angelos Amditis [319], manager of the EU-funded project "Adaptive Integrated Driver-Vehicle Interface" (AIDE) warns: "There is a real risk the driver will become overwhelmed as the number of in-car systems multiply. There are so many potential demands on driver attention from these new systems that they could prove distracting." Such information overload has the potential to result in a high cognitive workload, which subsequently reduces situational awareness and lowers driving performance [147].

A second issue is the complexity of the dashboard (or "cockpit"), which increases the likelihood of errors by the driver (either through spontaneous failure or design errors) [317, p. 249]. Tango *et al.* [320] presented a design method for Driver-Vehicle Interfaces, starting with a user needs analysis in order to ensure clear recognition of user and system requirements.

9.1 Alternatives Supporting the Driver

There are a number of options in order to compensate for the problems of cognitive overload and driver distraction, originating in the number and complexity of ADAS. The most powerful one would fully employ drivers' vital context, as presented in Fig. 9.3 (boldface sensor and/or actuator labels indicate technology utilized in the scope of this work). Of course this would necessitate a brand-new design of the vehicular controls and all interaction systems available today.

In a first step and as proposed within this research work, the sense of *touch* as an additional sensory modality would be used in the vehicle (integrated into the seat and backrest of the driver's seat). Vibro-tactile voice coil elements are used to generate vibrations (output channel); force sensor array mats are responsible for the acquisition of static and dynamic sitting postures of the driving person (input channel). An in-depth investigation on vibro-tactile interfaces is given in the next section.

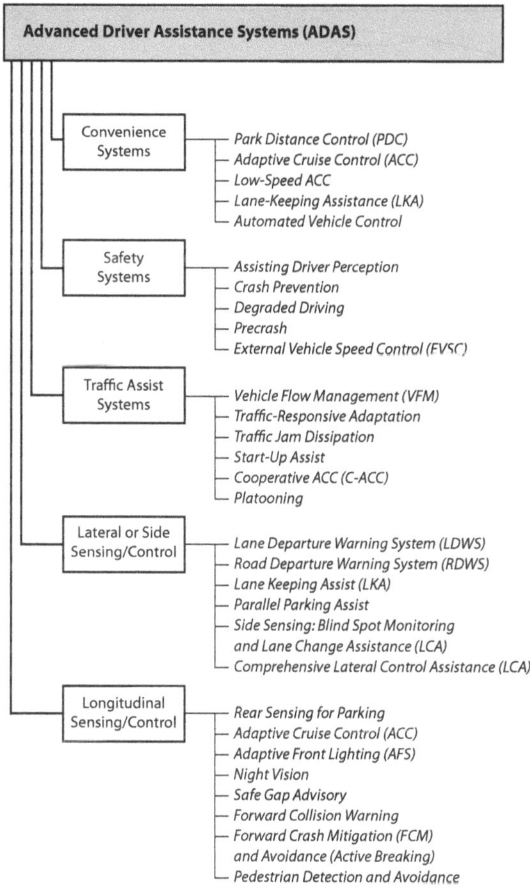

Fig. 9.2: Overview of Advanced Driver Assistance Systems (ADAS) (adapted from Bishop [9], van der Heijden *et al.* [10]).

10 Vibro-Tactile Interfaces

Innumerable studies have addressed performance analysis and impacts of using the sensory modalities *vision* and *hearing* in vehicles. Both of these communication channels have potential, but also drawbacks. Tactile feedback as a promising additional source of information is still underused in vehicles today [321, p. 9] and thus, worthwhile to investigate. Suzuki *et al.* [322] found, for instance, that *vibro-tactile* feedback on the steering wheel was effective for warning of lane-departure situations, especially for ad-hoc notifications (where drivers did not know the

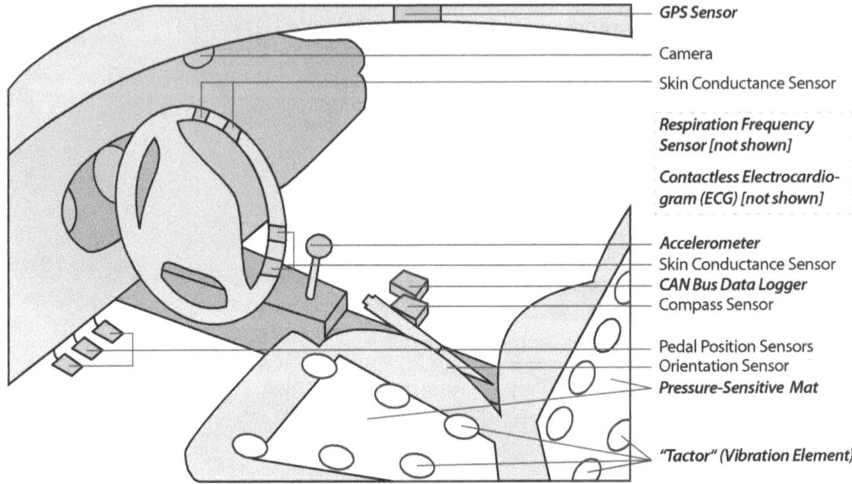

Fig. 9.3: Vital context in cars (adapted from Ferscha *et al.* [11] and extended by respiration frequency sensor, contactless ECG, pedal position sensors, and vibration elements).

meaning of warnings). Moreover, their experiments showed that it seems that many drivers have their own mental model for responding to a haptic stimulus [transmitted through the steering wheel]. Apart from using *haptics* in automotive applications as single information channel, there was also general concurrence with Wheatley and Hurwitz [163] who recommended using multimodal interfaces (a combination of *visual*, *auditory*, and *haptic* sensory channels) for safe and effective future Advanced Driver Assistance Systems.

10.1 Motivation

There is considerable interest with regard to the three major traditional senses *vision, hearing*, and *touch* in manipulating how these sensory modalities can improve human interaction with the external environment [124]. Especially haptic stimulation allows access to an important and often overlooked information channel in Human-Computer Interaction (HCI). Although vision accounts for the majority of sensory input (see pp. 6, 15, 27), the sense of touch is also a common information source. Tactile feedback makes it possible, for instance, to find the radio dial in a car without looking, and to know when fingers are correctly placed on a "QWERTY"-keyboard (because of the bumps on the "F" and "J"-keys). Akamatsu *et al.* [323] reported improvements in performance when using bimodal (audio and visual, tactile and visual) or trimodal (audio, tactile and visual) feedback, compared to visual feedback alone. Debus and others [324] showed that vibro-tactile feedback is useful for increasing user task performance (relative to visual feedback alone, vibro-tactile feedback reduced errors by 38%).

Considering the usage of vibro-tactile sensations from the perspective of losing the sense of touch provides interesting results. Furthermore, it shows that the usage of this sense is clearly undervalued today.

The total or partial loss of the sense of touch often cannot be adequately compensated for by the application of other sensory modalities and eventually results in a limited ability to perceive the environment or even to stand and/or move [130, p. 24]. As a consequence, it can be assumed that a vehicular application, when providing improper haptic feedback to a driver, might impair his/her performance in the same way as a major somesthetic loss would do. For example, it would be very difficult or even impossible to steer a car (using the brake pedal, throttle control, and/or clutch) with numb feet; but on the other hand, it would likewise be demanding for a driver if he/she received invalid or improper vibro-tactile feedback on a particular driving activity.

10.2 Types of Stimuli

The human tactile system provides unique and bidirectional communication between a person and his/her physical environment. Based on the underlying neural inputs, the tactile system can be divided into (*i*) cutaneous (employs receptors embedded in the skin), (*ii*) kinesthetic (uses receptors located in muscles, tendons, and joints), and (*iii*) haptic (combination of cutaneous and kinesthetic systems [325], and additionally incorporating an active exploration of the surroundings) subsystems (see Fig. 6.2 on p. 49 as well as the Paragraph "Definition of Terms" starting on p. 48).

For transmitting tactile information from the vehicle to the driver, several technical approaches are feasible. The nerves of the skin can be stimulated by electrical, thermal or mechanical methods [198].

The operational area of a vibro-tactile interface (available space for integration, existing power supplies, required vibration intensity and response time, spatial resolution of sensors – according to the receptor density in the skin, as shown in Fig. 10.2 on page 93, or restricted by unobstructed space, etc.) determines the choice of a specific system.

The following technologies should be possible either as stand-alone solutions or in combination with other systems (a more elaborate technical description is given, for instance, by Wünschmann *et al.* [326]).

10.2.1 Electro-Tactile

This device type directly excites the afferent nerve fibers responsible for normal, mechanical touch sensations in the skin by using electrical impulses from surface electrodes (according to Kurt Kaczmarek, [327]). They generate pressure or vibration-like sensations without using any mechanical actuator. Although this approach has been used to activate paralysed limbs, it only has potential in very limited application areas – the approach is highly invasive and not appropriate for the casual user [326]. The possibility of surgery and the potential liability in case of damage to the neuromuscular system further precludes this approach as an attractive alternative method of tactile/haptic interaction.

10.2.2 Heat (Thermal)

Heat or thermal stimulation of the skin in haptic interfaces can be provided using radiation (infrared and microwave), convection (air and liquid), conduction (thermo-electric heat pumps), or any combination of these. Thermal stimulation is a rather young research area; questions like "which temperature range offers the best resolution?" are still open (Jones *et al.* [328]).

10.2.3 Mechanical

Pneumatic
This type of mechanical stimulation involves using air jets, air pockets, or air rings. Pneumatic devices tend to have low bandwidth [329]. Users may eventually experience muscular fatigue, reducing their ability to sense. Furthermore, pneumatic systems cannot operate silently because of intaking and deflating air [330].

Piezo-Electric Crystals
These materials can be deformed specifically by applying electrical voltage. The amount of deformation is very low (1 to 4 ‰ of material length); when arranging the material in layers (sandwich) the proportion for deformation can be increased. Piezo-electric actuators have a fast response time (ms-resolution) and operate with low power consumption, but – on the other hand – require a high operating voltage of up to $1,000V$ DC [331, pp. 1–28], which is not always easy to generate and is prohibited in applications involving humans (International Electrotechnical Commission standard IEC 364-4-41 and German standard DIN VDE 0100-410[48] specify a maximum allowed voltage of $25V$ AC or $60V$ DC for direct contact).

Vibro-Tactile
Vibro-tactile stimulation involves using (i) blunt pins, (ii) unbalanced or eccentric mass motors, or (iii) voice coils to generate vibration. Vibro-tactile devices can be built very small, with a high vibration bandwidth. On activation, the moving part oscillates at a certain frequency

[48] http://www.vde-verlag.de/data/normen.php?action=normdetail&vertriebsnr=0100120&quicksearch=1&loc=de, last retrieved August 31, 2009.

(optimum stimulation is reached at the peak sensitivity of mechanoreceptors, which is around $250Hz$ for Pacinian corpuscles). Such devies are often the best way to address the somatic senses of a user (driver).

System Selection

According to early usability studies and a requirements analysis, demands associated with using vibro-tactile feedback in the automotive domain have been determined and are regulated by the following parameters: (i) unobtrusive integration into the driver seat, (ii) silent operation, (iii) fastest response times, (iv) high vibration intensity, (v) fail-safe and continuous operation, (vi) economic production, and (vii) ability to provide high stroke (vibrations had to run through the seat cover via the driver's clothes to the person's skin).

In accordance with these demands and considering the advantages and drawbacks of other available technologies, vibro-tactile devices are best suited to meet the requirements for both input via sensors and output via actuators in vehicles.

10.3 Stimulation via the Skin

The skin has, aside from other functions, the capability to perceive the environment with several sensors, such as pressure, vibration, temperature, or electrical voltage sensors [198]. Each of these sensation classes can be observed with one or more sensory organs, located in different layers of the skin and also in diverse regions of the body, and which are sensitive to specific modalities. The skin includes seven kinds of mechanoreceptors (sensitive to pressure, vibration, and slip) [199], two different thermoreceptors (sensitive to changes in temperature), four kinds of nocioceptors (responsible for pain), and three types of proprioceptors (sensitive to position and movement).

10.3.1 Cutaneous Mechanoreceptors

It is necessary to understand the characteristics of the cutaneous mechanoreceptors before a tactile display is designed. Fig. 10.1 outlines functional features of the four cutaneous mechanoreceptors (Ruffini corpuscles, Merkel discs, Meissner corpuscles, and Pacinian corpuscles), relevant for vibro-tactile stimulation (adapted and extended from [199], [332], [333], [334], and [335]). Each mechanoreceptive system consists of a mechanoreceptive receptor and an afferent neuron. These afferent neuron types comprise slowly adapting type 1 (SA1) afferents that end in Merkel cells, rapidly adapting (RA) afferents that end in Meissner corpuscles, Pacinian afferents that end in Pacinian corpuscles, and slowly adapting type 2 (SA2) afferents that terminate in Ruffini corpuscles [195, pp. 11–18].

Other Types of Mechanoreceptors

Apart from cutaneous mechanoreceptors, there are a number of other mechanoreceptors, including receptors in the hairy skin, hair cells (receptors in the vestibular system in the inner ear), or hair follicle receptors (sense the position of hairs). These types of receptors are inappropriate for the mode of stimulation required here (they are insensitive to pressure or vibration, but are sensitive to stroking, skin stretch, or position and movement) [336], [337], [338, p. 11], [198, p. 2], [339].

	Features of Mechanoreceptive Systems			
	MEISSNER CORPUSCLES	PACINIAN CORPUSCLES	MERKEL DISCS	RUFFINI ENDINGS
RATE OF ADAPTATION	Rapid	Rapid	Slow	Slow
LOCATION	Superficial dermis	Dermis and subcutaneous	Basal epidermis	Dermis subcutaneous
MEAN RECEPTIVE AREA	13mm²	101mm²	11mm²	59mm²
SPATIAL RESOLUTION	Poor	Very poor	Good	Fair
SENSORY UNITS	43%	13%	25%	19%
RESPONSE FREQUENCY	10 – 200Hz	50 – 1,000Hz	0.3 – 100Hz	0.4 – 100Hz
MIN. THRESHOLD FREQU.	40Hz	200 – 250Hz	50Hz	50Hz
SENSITIVE TO TEMPERATURE	No	Yes	Yes	Yes, at > 100Hz
SPATIAL SUMMATION	Yes	No	No	Unknown
TEMPORAL SUMMATION	Yes	No	No	Yes
PHYSICAL PARAMETERS	Skin curvature, sensed velocity, flutter, slip pressure	Vibration, slip, local shape acceleration	Skin curvature, local shape,	Skin stretch, local force

Fig. 10.1: Functional features of the four cutaneous mechanoreceptors, relevant for pressure/vibration sensation (the dashed frame indicates the type of receptor used).

All of the four cutaneous mechanoreceptors (Fig. 10.1) facilitate the sense of touch, each kind of receptor with its own characteristics for detecting and responding to stimuli. Stimulation in a vehicle, for instance, is provided by vibrations from a vibro-tactile display integrated in the vehicle seat, the steering wheel, or the handbrake. The tactile sensation perceived overall results in some respects from the superimposed afferent input of all of these four types of receptors [195, p. 9].

(*i*) **Ruffini Corpuscles:** The Ruffini corpuscles appear to play a substantial role in the perception of the direction of motion of a handled object and in the perception of shape and position. The SA2 (slowly adapting) afferents of the Ruffini corpuscle innervate the skin less densely than the other types of afferents (SA1 or RA) and their receptive fields are about five times larger. Thus, SA2 afferents have poor spatial resolution [340, p. 325]. Moreover, these afferents are less sensitive to skin indentation, but more sensitive to skin stretch than SA1 afferents [195, pp. 16].

(ii) **Merkel Receptors:** The Merkel cell or Merkel disk, a special cell type in the basal layer of the epidermis, is the most important structure primarily responsible for the sense of touch (in particular, for low frequency vibratory input in the range $0.3 - 3Hz$ [340]), tactile form and texture [195, pp. 12]. Merkel receptors are distributed across the body with about 25% coverage, and they are categorized as SA1 receptors, adapting slowly to stimuli. Merkel cells have a high sensitivity to points, edges and curvatures and they have a spatial resolution of only about $0.5mm$, overriding their receptive field diameter of $2 - 3mm$. SA1 afferents are at least ten times more sensitive to dynamic than to static stimuli, which indicates that for tactile input systems, which are in constant contact with the skin, stimulation of SA1 type receptors would not be the best application domain.

(iii) **Meissner Corpuscles:** Adjacent to the Merkel disk are the Meissner corpuscles, which are relatively large cell assemblies in the dermal ridges, located just below the epidermis, and comprising, for example, over 40% of the tactile receptors in the hand. Meissner corpuscles primarily operate as velocity detectors in the frequency range of $3 - 40Hz$ [340], providing feedback for grip and grasping functions, and requiring a minimum force for meaningful reception. Meissner afferents innervate the skin even more densely than the SA1 afferents of the Merkel cell. They are insensitive to static skin deformation, but are four times more sensitive to dynamic skin deformation than SA1 afferents. RA afferents respond to stimuli over their entire receptive fields ($3 - 5mm$) with relative uniformity and, therefore, resolve spatial details poorly. Some studies are underway using EAI[49] low frequency tactors to explore communication efficiency using Meissner corpuscles [195, pp. 14].

(iv) **Pacinian Corpuscles:** The Pacinian receptor is the largest and best understood type of the skin receptors, but is also the least common. The Pacinian afferents terminate in single corpuscles and furthermore, they are the deepest of the cutaneous mechanoreceptors and cover about 13% of the body. Pacinian afferents adapt very quickly and have almost no spatial resolution due to their deep location and their extreme sensitivity. The Pacinian corpuscles are mostly involved in high-frequency vibration detection [340, pp. 325] in the range between $50Hz$ and $1,000Hz$, with a maximum sensitivity in the frequency range of $200Hz$ to $250Hz$, and they serve as

[49] Engineering Acoustics, Inc., http://www.eaiinfo.com/TactorProducts.htm, last retrieved August 30, 2009.

acceleration detectors and vibration sensors. Pacinian corpuscles are categorized as the fastest adapting class of mechanoreceptors, RA1 (the effect of stimuli decays rapidly after onset). Pacinian corpuscles discharge only once per stimulus application; hence they are not sensitive to constant pressure [341].

Summary

Tactile perception results as the sum of the functions of these four mechanoreceptive afferent systems and it is not possible to pinpoing one particular afferent system as the best for any particular aspect of palpation. Neurophysiological studies have revealed the specialization in structure and function of these receptor systems and have demonstrated that each receptor system is specialized in conveying a particular type of information, such as *Merkel cells* for spatial resolution and tactile form, *Meissner afferents* for grip control, *Ruffini corpuscles* for skin stretch, and *Pacinian afferents* for vibratory stimuli – as tactile information presentation in this work should by provided by vibrations, the last named *Pacinian corpuscles* would be the best type for haptic stimulation. Bark *et al.* [342] also indicated that a vibratory stimulation applied to the skin innervates the class of fast-acting mechanoreceptors, namely *Pacinian* and *Meissner corpuscles*.

With regard to receptor properties in general and results from previous studies regarding vibro-tactile stimulation in particular, it seems that *Pacinian corpuscles* are best suited for tactile information presentation. This finding is confirmed, e. g., by Toney *et al.* [199, p. 36], who stated that most of the common tactor systems are designed to exploit the characteristics of *Pacinian corpuscles* because they are the class of mechanoreceptors most responsible for detecting vibro-tactile stimulations.

Previous work in tactile information presentation and perception, for instance in Cholewiak *et al.* [343], Brown [344, pp. 8–22], Myles and Binseel [345], Chouvardas *et al.* [198], Enriquez *et al.* [346], Hayward *et al.* [347], and Robles-De-La-Torre [130], has already investigated parameters such as tactor size and shape, vibration frequency and threshold, relaxation time, etc. for optimal stimulation of the body (or more specifically the mechanoreceptors or Pacinian corpuscles PC). Their results have been taken as initial points for further adaptation (the application domain "vehicle" requires unconsidered constraints, such as timing issues, distraction from the primary driving task, placement of actuators, etc.). Furthermore, as it has been substantiated that haptic stimulation depends on age[50] and/or gender, this factor needs to be considered when accounting for wide-ranging applicability. It has to be noted that Pacinian corpuscles, in particular, are afflicted by age while other receptor-types are less susceptible to age shifts.

[50]For details see the paragraph "Age Dependency:" in the "Simulating Real-Driving Performance" experiment on p. 162 or the corresponding part in Appendix D starting on p. 227.

10.4 Alphabets and Vibro-Tactile Patterns

A universal and broad application of vibro-tactile feedback in the automotive domain requires the definition of a vibro-tactile language or a haptic alphabet (the employment of vibro-tactile stimulation is also imaginable in other application domains; however, this probably would necessitate a re-definition of the tactile language, perhaps with other "characters"). Any vibro-tactile pattern applicable for the specific stimulation of a user and/or driver (e. g. with a minimum required awareness level) is defined in the "haptic alphabet", a pool of vibro-tactile signs or symbols.

The composition as well as the complexity of this alphabet should be oriented towards other, well-established alphabets (particularly situated in the "world of touch", such as the Braille alphabet, Tadoma, the Vibratese language, or even Fingerspelling) and their attempts to identify, define, and classify characters or words on different levels of importance (for instance, fingerspelling uses easy-to-show symbols for important and frequently occurring words, the Braille alphabet uses easy patterns for often-used characters, i. e. vowels, and defines abbreviations for the most common words like "and", "the", etc. as shown in Fig. 10.3).

Properties, advantages and drawbacks of these alphabets have been analyzed for reflecting the definition of a universal haptic alphabet or a language based on the sense of touch. A short overview of their basic elements is given in Appendix E: "Alphabets Related to Touch" starting on p. 230.

Threshold Distances for Different Parts of the Human Body

BODY PART	THRESHOLD DISTANCE	BODY PART	THRESHOLD DISTANCE
Fingers	2-3 mm	Upper Lip	5 mm
Cheek	6 mm	Nose	7 mm
Palm	10 mm	Forehead	15 mm
Foot	20 mm	Belly	30 mm
Forearm	35 mm	Upper Arm	39 mm
Back	39 mm	Shoulder	41 mm
Thigh	42 mm	Calf	45 mm

Fig. 10.2: Results of two-point discrimination threshold experiments, carried out by Kandal *et al.* [12], Gallace *et al.* [13], and Gibson [14], [15].

It is important to note that the definition of a haptic alphabet is not as easy as the definition of other touch-related alphabets, as, for instance, the number of vibro-tactile elements to present characters or words (*tactograms*) cannot be restricted. Depending on the application, the field of operation, or the placement of tactors on the body, more or fewer tactors could (and should) be used.

10.4.1 Tactograms – Related Application

The term *tactogram*, not to be confused with *actogram*[51], is not exclusively reserved for describing the dynamic activation of vibro-tactile elements for well-defined feedback. It is also used in fields ranging from marine sensory biology to business/management.

(*i*) Tactile detection in manatees: Manatees are unique marine mammals in that they have tactile hairs distributed over their massive girth. Joseph Gaspard *et al.* (University of Florida) investigates the sensitivity of the tactile hair system (and called this *tactogram*) and how manatees use this sense [348].

(*ii*) Virokannas [349] investigated the Vibration Perception Threshold (VPT) at six frequencies, applied in hand-arm vibration. Beside others, VPTs in the work of Virokannas have been assessed using a curve consisting of various frequencies, called a "vibrogram" (Lundborg *et al.*, [350]) or "tactogram" (Brammer *et al.*, [351]).

(*iii*) Gordios Retail Innovation A/S, Efficient Consumer Response (ECR) Europe [352] and ECR Europe Category Management Best Practices Report [353, p. 74] use the term "tactogram" for describing business planning processes (in-plane with strategies, approval processes, decisions, assigning responsibilities, scheduling, etc.). The Category Plan Implementation develops a specific implementation schedule and assigns responsibilites for completing all tactical actions shown under the category "tactograms".

(*iv*) Brammer *et al.* [354] (US Patent 5,433,211) described a method for identifying vibro-tactile perception thresholds of tactile nerve endings at a skin site of a subject to assess sensory change in sensory nerve function. Measurements where performed at six specific frequencies, chosen so that each mechanoreceptor population mediated the threshold at two frequencies. By introducing reference mean threshold values from a medically screened population, free from signs, symptoms or history of peripheral neuropathy, the thresholds determined at an individual's fingertip may be expressed as a "tactogram". This graphical representation of the data displays the change in the mechanoreceptor-specific threshold from the reference mean threshold value for normal hands at that frequency. A tactogram thus

[51] An actogram is a type of graph or chart commonly used in circadian research to plot activity (present or absent) against time, http://www.hhmi.org/biointeractive/museum/exhibit00/glossary.html, last retrieved August 30, 2009.

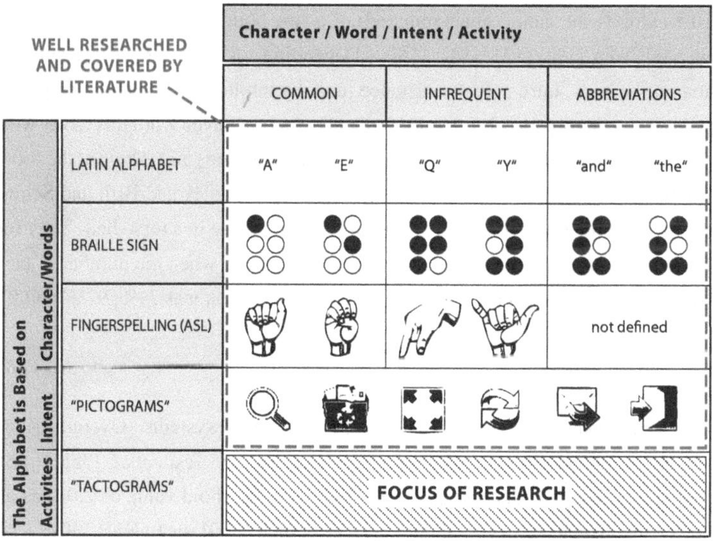

WELL RESEARCHED AND COVERED BY LITERATURE

The Alphabet is Based on		Character / Word / Intent / Activity					
		COMMON		INFREQUENT		ABBREVIATIONS	
Character/Words	LATIN ALPHABET	"A"	"E"	"Q"	"Y"	"and"	"the"
Character/Words	BRAILLE SIGN						
Character/Words	FINGERSPELLING (ASL)					not defined	
Intent	"PICTOGRAMS"						
Activites	"TACTOGRAMS"	FOCUS OF RESEARCH					

Fig. 10.3: Comparison of different alphabets (alternative forms of input) and their representation of typical characters, words, or activities (intent).

> expresses the acuity, relative to normal hands, of three populations of specialized nerve endings involved in the sense of touch, namely the SA1 (slowly adapting, type I), RA1, and RA2 (rapidly adapting, type I and II) populations.

The application of the term *tactogram* within the scope of this work does not overlap or intersect with these definitions of usage, and thus can be regarded as unique.

10.4.2 Parameters of Vibro-tactile Stimulation

The next paragraphs deal with the issue of tactor activation, primarily utilized in on back and bottom stimulation in a car seat.

Actuator Distance

It is a fact that receptor density is not uniformly distributed around the body (see Gallace *et al.* [13], or Gibson [14], [15]), and that different types of (mechano)receptors are not existent all over the body. Considering the different requirements for vibro-tactile systems in the automotive domain (such as rapid adaptation, physically responding to vibrations, spatial resolution), Pacinian corpuscles are best qualified as carriers for haptic feedback (see "Cutaneous Mechanoreceptors" on p. 89.).

Fig. 10.2 exhibits the data (values rounded) of a two-point discrimination threshold experiment (adapted from [12], [13], [14], [15]). Depending on the region or location of tactile stimulation on the body, intra-actuator distance (and, by implication, number and position, too) for an optimal presentation of vibro-tactile feedback can be derived from this table, which is for instance at least 39*mm* at the back (e. g. in a vibro-tactile car seat) and 20*mm* at the foot in order to be distinguished in each case as two different elements. Cholewiak, Brill and Schwab [343, p. 979] investigated the number and placement of tactors for use in a torso belt. They found that the localization performance of vibro-tactile stimuli improved when the number of tactors was reduced from twelve to six – the overall accuracy of localization was 74%, 92%, and 97% for a belt with 12, 8, and 6 tactors.

Response Time

Several researchers have investigated response times for HCI systems. Overall recommendation for the maximum response time is two seconds or less, see Testa *et al.* [355, p. 63], Miller [356], or Shneiderman and Plaisant [27, Chap. 11]. The threshold value of 2.0*s* has also been confirmed for the visual sensory modality, e. g. in French [300], or Labiale [301] – they indicated that times greater than 2.0*s* for visual tasks are unsafe and unacceptable.

Ben Shneiderman [357] presented an extensive study on response times and their influence on Human-Computer Interaction performance. His most important findings, relevant for this work, include (*i*) results of a large study indicate that 2 seconds was generally an acceptable response time, (*ii*) in the absence of constraints, such as technical feasibility, people will eventually force response times to well under 1 second, (*iii*) people will work faster as they gain experience with a command – so it may be useful to allow people to set their own speed of interaction, (*iv*) more rapid interactions (less than 1 second) are normally preferred and can increase productivity; this is line with the statement that "people can adapt to or work with longer response times, but they are generally dissatisfied with doing so".

Due to the computation power of today's computer systems or Electronic Control Units (ECUs) and the bandwidth of bus systems (MOST, FlexRay, etc.), the time required for a response – mainly determined by collecting and processing of sensor data – often can be neglected.

Activation Time (AT)

Taking the results from the "Response Time" section above, the maximum cycle time for one vibro-tactile pattern (tactogram) should not exceed the upper limit of 2 seconds in order to avoid annoyance and performance loss. Particularly in the automotive domain, situations change often and require the driver to react fast (and spontaneously) – this would demand fast and short vibro-tactile notifications (with respect to the delay of response and stimulation length) of important information, too.

Considering the recommendations for response times (particularly that of Shneiderman), it would be ideal if the cycle time of a vibro-tactile pattern was in the range of one second or below. This value has been confirmed by experimentally determined timing parameters, e. g. Scott *et al.* [358] experimented with driver (brake) reaction times for visual, auditory and tactile feedback with a cycle time for vibro-tactile stimulation of $1.0s$ ($200ms$ activation, $800ms$ pause), and van Erp and van Veen [168] investigated visual and tactile navigation displays in vehicles. Depending on the distance to the next waypoint, they varied the inter stimulus interval and the number of bursts (with a fixed burst duration of $60ms$). The maximum cycle length was $1.156s$ (6 bursts of $60ms$, disconnected by breaks of 264, 212, 160, 108, and $52ms$).

Vibration Amplitude (VA)

The Pacinian corpuscles react to vibrations within a large frequency bandwidth [359], [360] (see the section "Vibration Frequency (VF)" for further details). As it is established that the perceived intensity of vibration varies as a function of both frequency and amplitude (shown in Fig. 11.4), the intensity of a stimulus is most often coded by variation of frequency only [361] (e. g. because of the ease of implementation, particularly when using low-cost cell phone motors). However, if the amplitude of a vibration is changed while the frequency is kept constant, a noticeable change in the stimulus would occur (as expected) [362, p. 9]. In this context, the term "threshold" is the smallest stimulus intensity or change in intensity that is noticeable (a stimulus lower than the threshold level does not activate nerve fibers to fire, and thus generates no action).

Visell [360] indicates that the threshold of Pacinian corpuscles (on the hand) is between $3.0\mu m$ and $20.0\mu m$ with a mean of $9.2\mu m$, Kaczmarek *et al.* [363, p. 3] refers to research results from Gescheider and Verrillo, which confirm that the skin's sensation threshold for the Pacinian type of receptors is frequency dependent, with best sensitivity achieved at $250Hz$ and a stimulation area of $5.0cm^2$ (threshold $0.16\mu m$). Furthermore, Gescheider and Verrillo [364] found that the threshold amplitude for vibro-tactile stimulation increases after a strong conditioning stimulus for frequencies between $10Hz$ and $250Hz$ (a 10-minute stimulus of $6dB$ over threshold raises the sensation threshold amplitude by $2dB$, while a $40dB$ stimulus raises the threshold by $20dB$). Verrillo [365] discovered that the perception of Pacinian corpuscles due to stimulation summates over time (the vibro-tactile threshold for a $250Hz$, $2.9cm^2$ stimulus falls by $12dB$ as stimulus time increases from $10ms$ to $1,000ms$).

Vibration Frequency (VF)

In general, Pacinian corpuscles are sensitive to vibrations in the range between $50Hz$ and $300Hz$ ($1,000Hz$), with a maximum perception intensity (resonant frequency) at around $250Hz$ [342], [340], [360, p. 4], [359]. According to Cholewiak *et al.* [366], the frequency value actually used for haptic stimulation affects the ability to detect and locate tactile information. They suggested using stimulation frequencies in the range between $80Hz$ and $250Hz$. Fitch *et al.* [367] conducted experiments for collision or crash avoidance and found that the optimal stimulation frequency range is between $100Hz$ and $140Hz$ (frequencies below $100Hz$ were omitted in order to ensure a clear separation between vibrations from the road – which typically occur at $50Hz$ or below – and the actual tactile feedback).

Bark *et al.* [342] found that the afferent fibers activating the Pacinian corpuscles fire at a rate proportional to the frequency of the stimulus (as opposed to other mechanoreceptors which typically code stimulus amplitude with firing rate).

10.4.3 General Remarks

A vibro-tactile alphabet, defining patterns for specific stimulations, should not operate on character level (as known for common alphabets, and exemplarily depicted in Fig. E.1, p. 231), but on *activity* level as shown in the Figures 10.3 and 10.5. In addition, *commands* or *activities* for a haptic alphabet are required to be application or domain specific (there are more notification demands in the vehicular domain compared to e. g. control stations in industrial plants or office environments).

10.5 Tactograms

In order to be able to present vibro-tactile patterns visually (and humans are best qualified to interpret visual representations), a display format (based on other representations e. g. for printed/sheet music) needs to be defined. A coding schema referred to as *tactograms*, as shown in Fig. 10.4, has been introduced, where the mapping of all sorts of variation parameters for representing one specific tactor element is considered as follows: The x-axis indicates the activation time (AT), the vibration amplitude (VA) is assigned to the y-axis, and the fill gradient of the rectangle, which is attributed to the z-axis, is an indicator of the vibration frequency (VF), ranging from white (lowest value, 0%) to black (highest possible value, 100%). The values in all dimensions are scaled to the range 0 [minimum value] ... 1 [maximum value].

This form of illustration is universal; the border values for each dimension (0, 1) must be assigned based on physically or experimentally evaluated limitations of the actual system (furthermore it is not certain that the pathway between 0 and 1 is always linear – most often this would actually not be the case).

For instance, when thinking of a tactile navigation system, the remaining distance (determined by, e. g., a GPS sensor) to the next "activity" (waypoint) could be either mapped to the activation time, the vibration frequency, the vibration amplitude (intensity or attenuation), or any combination of these parameters.

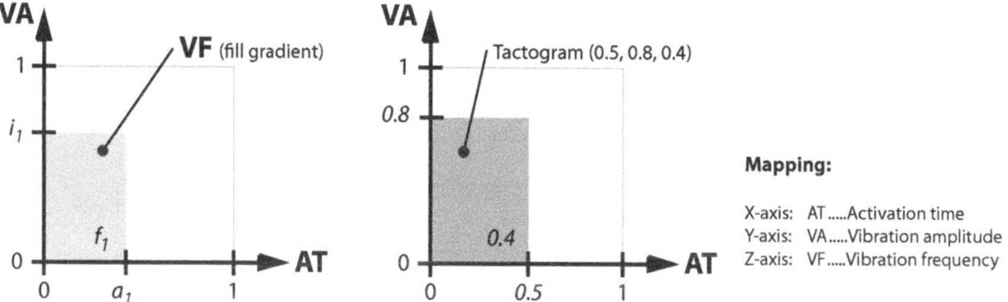

Fig. 10.4: A *tactogram* is defined unambiguously by a 3D tactor pattern representation. The variation parameters for a specific tactor element are activation time, vibration intensity, and vibration frequency.

10.5.1 Related Work

Krausman *et al.* [368, pp. 5–7] experimented with six different (turn right, turn left, move forward, turn around, stop at next obstacle, stand in place) vibro-tactile patterns, presented alternatively on two tactile display types (4x4 tactor array as back display, 8 tactor belt display). The visual representation of patterns was specified in a similar configuration to that of the present work. The researchers reported a frequency of correct mappings between 80.4% (stop) and 93.3% (turn left) for the different patterns (the type of tactile display – belt or back – did not affect detection and identification of the tactile patterns), evaluated as the driver maneuvered through a mobility-portability course. The high identification rate for all patterns demonstrates the ability of tactors to communicate several different meaningful patterns to a person.

Jones *et al.* [201] performed a similar study using a 4x4 array of vibrating motors as tactile display. They reported that test participants were able to interpret eight tactile patterns for navigation aid with almost perfect accuracy. Furthermore, they found in modified experimental settings that (*i*) there is no difference in the performance between two different types of vibration motors, and (*ii*) vibro-tactile stimulation on the back performs better than stimulation on the forearm.

Both of these studies affirm the efforts in this field of research and encourage further endeavors in this area.

10.5.2 Multi-Tactor Systems

Real tactile notification systems would consist of more than one vibro-tactile element (*tactor*), and thus require an extension of the representation introduced as *tactograms* (Fig. 10.4).

Static View

First of all, a format for defining the static system design is required, showing the number and placement of tactors in a specific application, and optionally a mapping of vibro-tactile elements utilized for the notification of specific actions required.

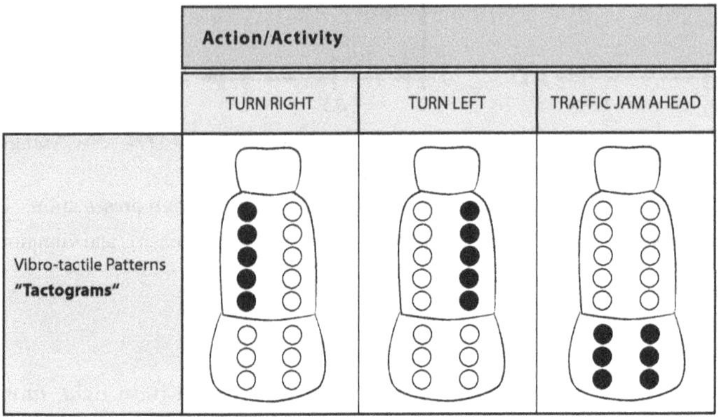

	Action/Activity		
	TURN RIGHT	TURN LEFT	TRAFFIC JAM AHEAD
Vibro-tactile Patterns **"Tactograms"**			

Fig. 10.5: Static view of a multi-tactor system defining *tactograms* for three activities.

Fig. 10.5 depicts one possible system design for application in the driver seat of a vehicle notifying about the three approaching situations "Turn Right", "Turn Left", and "Traffic Jam Ahead". Each circle represents one vibro-tactile element; the black filled ones are the tactors employed for a certain activity (e. g. the five tactors on the right side of the back of the seat for the activity "Turn Right").

Dynamic View: 6-Tuple Notation

For a specification of dynamic activation traces of vibro-tactile systems, the 3-tuple notation (AT, VA, AF) describing the behavior of one tactor element and the static view of a multi-tactor system are combined to display the dynamic sequence of tactor activation (see Fig. 10.6).

A 6-tuple notation (T, X, Y, AT, VA, VF) is used for a formal definition of the vibro-tactile dynamics. The first three parameters are the actual point in time (common time base) and the x and y coordinates of the tactor in the considered tactile notification system. The remaining three parameters are taken from the 3-tuple notation describing one-tactor systems (these are the variation parameters activation time, vibration amplitude and vibration frequency).

Findings on the variation of stimulation parameters, investigated in various pilot studies, are (*i*) the higher the vibration amplitude, the higher the achieved level of attention LOA, (*ii*) a change in the vibration frequency is associated with an adjustment in the level of attention (highest level of attention at the nominal level of about $250Hz$, decreasing (LOA) above or

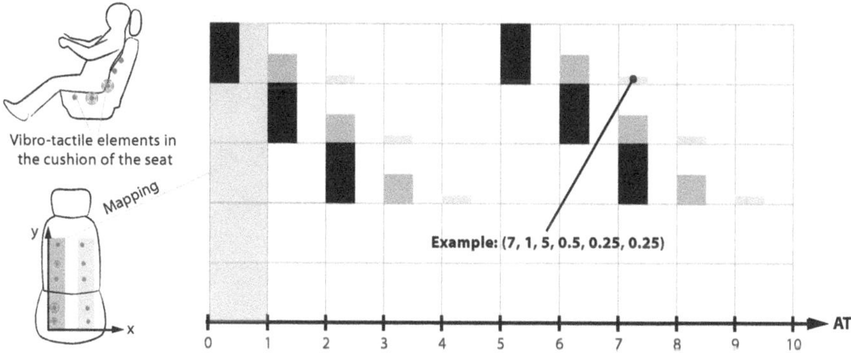

Fig. 10.6: The graph on the right side shows the dynamic flow of tactor activation for a multi-tactor system consisting of 2 x 5 elements. The tactile *image* is called a *tactogram* (or *vibration carpet*); individual elements are specified with a 6-tuple.

below this nominal frequency), (*iii*) changing the pulse-pause ratio directs to a higher level of attention when the ratio between pulse and pause increases (=lower pause time) and vice versa, (*iv*) the longer the activation time, the higher the level of attention of the user (but it must be noted that human beings adapt to durable stimulation – and in that case the level of attention tends toward zero), and (*v*) the rhythm of vibrations influences the generated level of attention (harmonic or disharmonic patterns cause lower or higher attention levels).

10.5.3 Level of Attention (LOA)

This specific form of vibro-tactile representation and stimulation creates the possibility of investigating patterns (tactograms) for generating a certain level of attention (LOA) on the influenced person, as outlined in Fig. 10.7. As opposed to other studies on vibro-tactile stimulation, this approach is unexplored in haptic notification systems, and would offer great potential for pleasant vibro-tactile interfaces[52].

Findings of experiments with different levels of attention are, amongst others, (*i*) the definition of a set of basic patterns allows a global "vibro-tactile language" to be defined (based not on letter or literal-level, but on activity or command-level specific for an application domain), (*ii*) each (type of) vibro-tactile pattern generates a predefined LOA and thus can be explicitly used for optimal notification of information based on its importance, and (*iii*) choosing and applying an appropriate tactogram (which, in this context, means a pattern raising the minimum required

[52]The general approach is not new, actually, and was already proposed by Matthews *et al.* [297] in 2004, however, only for visual displays.

LOA) for a specific notification requirement increases overall system performance (and reduces a driver's cognitive load).

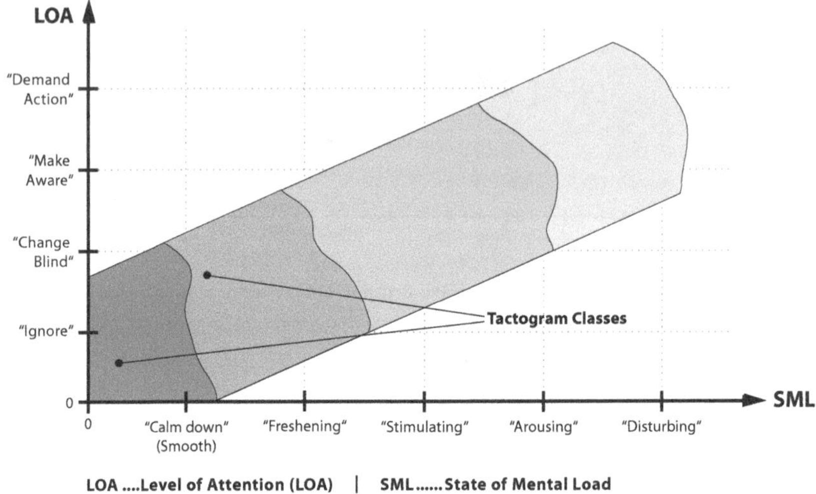

Fig. 10.7: Specific vibro-tactile patterns for controlling the user's *level of attention (LOA).*

Considering a specific notification requirement in the vehicle creates the need to choose a style or mode of stimulation correlating to the importance of the information. Therefore, a fixed number of importance levels needs to be defined and each information demand has to be assigned to one of these levels. According to the definition of Matthews *et al.* [297, p. 249], the term "notification level" (which refers to differences in information importance, too) is used from here on, instead of "importance level".

McCrickard *et al.* [40] defined the three levels of notification systems (i) "Interruption" (draw a user's attention), (ii) "Reaction" (rapid response to a stimulus), and (iii) "Comprehension" (remembering and sense-making). Matthews *et al.* named the five levels (ascending according to their importance) (i) "Ignore", (ii) "Change Blind", (iii) "Make Aware", (iv) "Interrupt", and (v) "Demand Action". Pousman and Stasko [369] suggest replacing Matthews' lowest level of notification ("Ignore") with "User Poll" in line with the fact that information of low importance is not displayed automatically, but is explicitly requested by the user.

In this work, the lowest three notification levels have been taken from the definition of Matthews *et al.* However, an interruption of the primary task of driving in the automotive domain generates a high cognitive load on the driver, and in succession leads to distractions – due to this fact, Matthews' $(iv)^{th}$ class ("Interrupt") has been merged with her $(v)^{th}$ ("Demand

Action") to a new level-(*iv*) class "Demand Action", finally leading to the following structure of notification levels:

(i) **"Ignore":** Represents a notification demand concerning unimportant information and should require no additional attention of the driver. A typical example of this class would be the feedback generated by an activated turn indicator.

(ii) **"Change Blind":** This class is assigned if a notification requires no immediate reaction by the driving person, but later action in any case, and is delivered to the user with a pattern causing a low level of attention (LOA). An example of this class would be a flashing bulb on the dashboard, informing the driver that the fuel level is down to the reserve.

(iii) **"Make Aware":** User feedback with this notification level requires immediate action, but would not cause serious danger if the response is time-delayed, as, for instance, the information "the filling level of the engine oil is below the minimum – stop the car!" (if the car is driven for another 5 or $10km$, nothing dangerous should happen). Notifications of this level are delivered to the user with a vibro-tactile pattern, provoking a medium level of attention (LOA) of the driver.

(iv) **"Demand Action":** This class of highest risk potential necessitates an immediate reaction by the driving person in order to prevent hazardous situations. Events of this notification level are critical enough to demand a user's full attention. One example would be the warning of a chain reaction pile-up just ahead, which would require an immediate emergency braking by the car driver.

These four stages of LOA (ignore, change blind, make aware, and demand action) can subsequently be used for generating a particular cognitive/mental load on the driver in connection with specific notification demands. The mapping, which is person specific, is not fixed and can be changed even during a ride ("at runtime"), depending on the actual environmental conditions, e. g. a notification of "low fuel" would be delivered with a higher level of attention on the highway than in a city (depending on the density of or distance between gas stations).

*"All truths are easy to understand once they are discovered;
the point is to discover them."
Galileo Galilei (1564 – 1642)
Italian astronomer, philosopher, and physicist*

Part IV

Methodology

For the investigation into the research hypotheses of this work, as presented in the section "Hypotheses and Research Questions" starting on p. 30, several experiments have been conducted. Basically, two categories of interfaces have been utilized, (*i*) a pressure sensing interface analyzing sitting postures built up from high-resolution sensor arrays integrated into the driver's seat and backrest, and (*ii*) a vibro-tactile notification interface constructed from electro-mechanical coil transducers, both embedded into the cushion of the seat. For improved accuracy, both the sensing and the (vibro-tactile) stimulation experiments have been conducted and evaluated separately.

The basic conditions of the individual experiments have been defined according to the analytical methods described in the following and well-known general limitations, such as, for instance, the difference threshold[53].

11 Analytical Methods

Collected data have been evaluated and analyzed with the methods indicated below, using different mathematical/statistical tools, e. g. Matlab, Weka (machine learning software), HTK (Hidden Markov Model toolkit), RapidMiner (formerly known as YALE – Yet Another Learning Environment), the Unscrambler (software package for multivariate analysis and experimental design), LIBSVM (Library for Support Vector Machines), and GHMM (General Hidden Markov Model library).

The numerical computing environment "Matlab©" (version 7.0 R14) has been identified as being best suited for data processing, interpretation, and charting for all types of conducted experiments, and thus has been used for the fina levaluation of data from all experiments.

The applied methods have been chosen in consideration of the required computation power (time) and the applicability (and availability) for real-time operation in vehicles.

11.1 Requirements and Technological Conditions

For investigating the applicability of the *sense of touch* (and the related terms *force*, *pressure*, and *vibration*) for both touch based input and vibro-tactile output, general requirements and conditions have to be defined prior to the experimental setup.

Based on the findings from (*i*) human perception and articulation, (*ii*) cognitive (over-)load, (*iii*) activity and notification demands, (*iv*) vibro-tactile interfaces, and (*v*) level of attention (LOA) (see Parts I to III of this work), the following technological conditions and requirements for a universal system architecture have been identified.

[53]Difference threshold is commonly known as Just Noticeable Difference (JND).

11.1.1 Force-Sensitive Input

(i) **Size:** The force-sensitive area should match both size and form of the vehicle seat and backrest as well as possible. The utilization of two sensor mats of about $50x50cm^2$ each would be preferred to the integration of one larger mat (flexibility of use).

(ii) **Sensor density:** Precise data acquisition necessitates a large number of accurate sensors, paired with a small intersensor distance. The maximum number of sensors per mat is restricted by physical and/or fabrication limits and is in the range of one sensor per cm^2 (48 by 48 = $2,304$ sensors on a square mat with $48.8cm$ side length, see p. 201).

(iii) **Range of weight:** A system using the weight of a driver, e. g. for person identification, has to be defined in a way to cover all persons possible as drivers (for example a range between $40kg$ and $180kg$).

(iv) **Update rate:** The refresh rate of sensor values has to be fast enough to detect all driving dynamics. Only in this case, accurate and reliable operation of an application using force-sensitive input can be guaranteed.

(v) **Material:** The force-sensitive mat should be robust, flexible and thin, so that it is not perceived as uncomfortable by persons sitting on it.

Note: A requirements analysis in more detail is given in Appendix A: "Hardware for Processing Pressure and Vibration" starting on p. 199.

11.1.2 Vibro-Tactile Output

(i) **Placement:** The utilized vibro-tactile elements have to be integrated into the cushion of a vehicle seat without generating any disturbance, thus have to be small (thin), light-weight, provide strong, localized sensation, and operate silently.

(ii) **Variation parameters:** The tactors should allow for independent variation of vibration frequency, vibration amplitude, activation or pulse-pause time, as well as individual control of each element in a larger system (without interdependencies).

(iii) **Range of vibration frequency, amplitude**: In this work, the Pacinian corpuscles are stimulated in order to deliver vibro-tactile information. Stimulation should be possible in the full range of response of that type of receptor (e. g. in a frequency range from $50Hz$ to $1,000Hz$, see Fig. 10.1).

(iv) **Number of tactors:** All evaluations are done on a seat in a real vehicle with vibro-tactile elements integrated into both the seat and backrest. The number of tactor elements required for optimal feedback can be calculated from the threshold distances of the relevant body parts (e. g. 39*mm* for the back, see Fig. 10.2).

(v) **Power supply:** The control interface and the vibration elements are operating on safety extra-low voltage (SELV) direct current (DC) in order to assure safety in case of failures.

11.1.3 Other Sensors

For the experiments, a number of supplementary sensors have been integrated for different purposes. They are partly required for evaluations (e. g. accelerometer data or vehicle speed), partly only recorded to additionally inspect the results (e. g. a visual examination using video cameras), or are even recorded with no actual disposition, for later investigations with a larger number of input dimensions or sensors (e. g. ECG device).

11.2 System Design

11.2.1 Posture Pattern Analysis

For the experiments on pressure sensing and sitting posture pattern analysis, data from the driving person have been acquired via two pressure-sensitive arrays, integrated into the driver's seat and backrest. Depending on the hardware utilized for sensing[54] and the selected update rate, more or fewer data per time unit need to be processed. As an example, an estimation of expected data is quoted for the system setting used in the first prototypes: The FSA mat system has the capability to deliver data in a two-dimensional array representation of 32 by 32 values (12*bit* resolution each [370]) at a refresh rate of up to 10*Hz*. This implies a maximum of more than 20,000 data points per second (see equation (11.1)), and requires – especially when operating in real-time applications with dynamic posture traces – robust and sophisticated methods for processing and analyzing.

$$\text{data}_{max} = 2\ [\text{mats}] \times (32 \times 32\ [\text{sensors}]) \times 10\ [\text{Hz}] = 20,480 \qquad (11.1)$$

For the first experiments on person (driver) identification using static postures[55], the feature vectors have been calculated from a weighted combination of several parameters. Experiments

[54] Two sensing systems of different manufacturers, one with 32 by 32 sensors and another with 48 by 48 sensing elements were used.

[55] The posture patterns used for the analysis were calculated from the arithmetic average of about 20 consecutive readings.

on activity recognition used dynamic patterns, acquired with a refresh interval of up to $10Hz$. The amount of data processed in one experiment pass was counted for the latter series with about 6.2×10^6 data points (or approximately $15MB$ quantity of data). For the ongoing experiments on dynamic posture evaluation and real-time applications with sensor arrays comprising both increased resolution and higher update rates, a significantly larger volume of data is expected, requiring methods for reducing the calculation complexity. Below, a short summary on feasible techniques[56] is given.

Corresponding experiments in this work making use of this kind of processing are "Supporting Implicit Human-to-Vehicle Interaction: Driver Identification from Sitting Postures" (p. 121) and "Intelligent Vehicle Handling: Steering and Body Postures While Cornering" (p. 138).

11.2.2 Vibro-Tactile Notifications

When interacting with the user by applying systematic pressure patterns with varying intensity on different parts of the body, it is essential to know the correlation between human response and physical stimulus. It is known that the human sensory modalities adapt their capacity according to environmental (or cognitive) conditions. For instance, for the visual sense, the eye has the capability to see in broad daylight and also in the dark [371] (with different types of receptors). In the process of adaptation to the dark, each eye adjusts according to the brightness level of the environment (from a high-luminance setting to a low-luminance setting).

This adaptation process is also valid for hearing, tasting, and the sense of touch – in view of this knowledge, adaptation of vibro-tactile stimulation in haptic interfaces definitively needs to be considered.

11.3 Eligible Methods for Pressure Sensing

11.3.1 Multivariate Data Analysis (MDA)

Factor analysis is mostly used (i) for *reducing* the number of variables or dimensions in large problem classes and (ii) for *classifying* variables by detecting relations among these variables. Multivariate techniques (MDA) could be used to improve the processing of pressure sensor readings (up to $1,024$ readings at a $10Hz$ update rate).

Principal Component Analysis (PCA)

Principal Component Analysis is the simplest of the eigenvector-based multivariate analysis techniques for reducing the number of variables in the original set by replacing them with a

[56]All of them are well-established in image or video-recognition applications; pattern recognition from sitting postures is a closely related scope of work

small number of *factors*, called Principal Components (PC), as a linear combination of the original variables. As the PCs are orthogonal to each other, there is no redundant information.

PCA is commonly used in image and video-processing (Slivosky and Tan [241], Smith [372]). As posture maps (or images) can be indicated as a special form of regular images (in the case of static evaluation) or videos (when analyzed dynamically), Principal Component Analysis should be a proper technique for processing data delivered via the pressure mats: For instance, the 32 by 32 sensor elements per mat can be interpreted as *pixels* in common images. The 12bit sensor values (0...4,095) therefore simply need to be converted into a matching *color schema*. Computational complexity can be estimated as follows: For a given image size (or number of pressure sensing elements) of $size = r(ow) \times c(olumn)$, the training set consists of points in an n-dimensional space ($n = r \times c$); therefore, the processing complexity is $O(n)$.

When inspecting sitting posture patterns, compact regions ("image objects") could be identified, caused for instance by high pressure in the pelvic bone regions or by the thighs on the seat. These smaller subspaces of the original image can be used for reducing training sets. Calculations in the lower-order subspace have the advantage of being more efficient and faster (as indicated e. g. in [373], [374, pp. 125], [372], or [375]). Tan [376], for instance, applied an image-processing algorithm based on PCA to reduce the sensing points from a sitting posture distribution map to a few potentially important features[57] and achieved a reduction of more than 99% of the variables with unchanged recognition accuracy.

11.3.2 Linear Discriminant Analysis (LDA)

Linear Discriminant Analysis (LDA) computes a linear predictor from two data sets with the aim to classify future observations (Fisher [377], McLachlan [378], Duda *et al.* [379], Friedman [380], Martinez and Kak [381], Goe [382], and Balakrshnama *et al.* [383]). LDA seems to be well suited as a method for dynamically analyzing sitting posture traces in order to predict a driver's future steering activities.

11.3.3 Hidden Markov Models (HMM)

Eigenvector-based methods, such as Principal Component Analysis (PCA), are best suited for determining static sitting postures by reducing the number of relevant sensing points. One step further, continuous tracking of posture patterns (in real-time) requires (i) indentification of static sitting postures, and (ii) modeling of the transition among them. One feasible technique for continuous posture determination (prediction) should be the utilization of Hidden Markov Models (HMM) [376].

[57] Features in sitting posture maps are, for instance, the position or the distance of the pelvic bones (highest pressure points), the total contact area, or the entire weight (pressure) applied to the mat.

Mota and Picard [243, p. 4] used a set of independent Hidden Markov Models (as described e. g. in Rabiner [384] or Seymore *et al.* [385]) for modeling sitting posture sequences. In their evaluations, they reported a recognition accuracy between 76.5% and 82.0%.

Comparing related activities in the recognition and analysis of dynamic human postures, the proposed approach of analyzing posture patterns, collected from pressure sensitive arrays on the seat and backrest, is a novelty, as most of today's systems operate on the basis of video processing methods (see for instance DeFigueiredo *et al.* [386]).

11.3.4 Other Methods

Apart from the established (more or less sophisticated) mathematical methods Bayes nets, Neuronal nets, Fuzzy Logic, machine learning techniques, etc. (for details see e. g. Bishop [9], Demuth *et al.* [387], Morari and Ricker [388], or the Matlab Handbook [389]), Slivosky *et al.* [241, p. 37] suggested using a subset of the extractable features from a sitting posture pattern image, such as, for instance, (*i*) the total force applied on the mat, (*ii*) the average pressure value, (*iii*) the maximum pressure value, (*iv*) the number of components, (*v*) other prominent associated areas in the pressure map, or (*vi*) the angle of divergence of the legs, or major/minor axis. All of these parameters can be easily detected and processed – in particular the last item (*vi*) seems to be of major interest in a rapid validation of whether a specific person is sitting on the seat or not.

11.3.5 Summary and Impact

Different techniques of multivariate analysis, each with its suggested field of application, allow for the observation of a large number of variables at the same time.

Experience has shown that Principal Component Analysis (PCA) is suitable for the processing and analysis of posture patterns from the sensor mats. Nevertheless, apart from the preferred method PCA, other techniques for dynamic posture evaluation are to be applied in order to compare their applicability and performance for particular experimental settings. It has already been shown that good results are obtainable with even simpler methods, as proposed for instance by Slivosky *et al.* [241] or shown in the experiments 1 and 3.

On-board real-time data processing (by a specific Electronic Control Unit) would benefit from reduced calculation complexity due to the lack of computation power.

11.4 Techniques for Vibro-Tactile Stimulation

11.4.1 Stimulus Detection

Miller [390] determined that people can rapidly recognize approximately seven (plus/minus two) "units" of information at a time (and hold them in short-term memory). This value has been debated in later research works and, in particular, for haptic information transmission Gallace *et al.* have shown that untrained people can only discriminate between four levels of vibro-tactile stimulation. Moreover, they reported that people's ability to detect stimuli presented simultaneously over the body surface decreases (substantially) when the number of tactors activated exceeds three [219], see also [220], [221].

A 2007 article of Gallace *et al.* [220] summarizes the research in the utilization of tactile stimuli on the body surface over the past half century. Substantial findings of this review are: (*i*) People are limited in counting the number of stimuli presented at any one time on their body, (*ii*) participants often fail to report the second of two targets in a sequence of events (both must be reported, the second follows shortly after the first), (*iii*) tactile interfaces might provide an additional means of stimulation for elderly individuals experiencing a reduction in sensory sensitivity in different sensory modalities, (*iv*) tactile sensitivity itself deteriorates with age (requires appropriate compensation), and (*v*) tactile interfaces are predestined to be used as add-ons in multisensor environments (where many sources of information are provided to a person at the same time).

There is evidence that people are poor at detecting significant changes between visual scenes (change blindness) or between auditory patterns (change deafness) [391, p. 300], but not those related touch. In the last years the interest for tactile interfaces has grown rapidly, and with the rising demand for interfaces based on the sense of touch it is becoming more and more important to understand perception limitations in order to overcome the disregard for this information channel. Gallace *et al.* [391] reported the existence of tactile perception limitations even for simple and infrequently changing vibro-tactile patterns. The tactile deficit appears of particular note for information transmission via tactile interfaces (e. g. type and amount of information, speed).

In a recent study, Gallace *et al.* reported that representations of tactile stimuli are stored in the cognitive system for a limited amount of time. Further results have been presented, for instance, by Gallace *et al.* ([13], [221], [392], and [393]), Hayward *et al.* ([394]), Ju *et al.* ([39], [137]), and Tan *et al.* ([395], [396]).

11.4.2 Discriminating Stimuli

The *Just Noticeable Difference (JND)* or *Difference Threshold* is defined as the minimum amount by which the intensity of a specific stimulus must be changed in order to produce a noticeable sensory experience (see Montag [397], Schiffman [398, p. 371]).

Weber's Law

Ernst H. Weber experimented with blindfolded test persons and variations in weight of held items and found that there is a linear correspondence between the weight of an item and its human perception. Based on these findings, *Weber's* law (1834) states that the ratio of the increment threshold JND to the background intensity is a constant [105, p. 288]. He experimentally determined the value of the *Weber* fraction [399] ($\frac{\Delta I}{I}$) for different senses: It is about 3% for the sense of touch, 1–2% for the sense of vision, and between 10% and 20% for the sense of taste [398, p. 15]. Applied to the automotive domain, when a person is driving at high speed and it is raining (=high level of noise on windshield and roof), then the driver must shout to be heard by co-passengers while a whisper works when the car is stopped and there is no environmental sound. When increment thresholds are measured against backgrounds of varied intensity, the thresholds increase in proportion to the background.

Weber's law can be applied to a variety of sensory modalities; the size of the *Weber* fraction varies across modalities but in all cases tends to be a constant within a specific modality (ΔI represents the difference threshold, I represents the initial stimulus intensity and k signifies that the proportion on the left side of the equation remains constant despite variations in the I-term):

$$\Delta I = k \times I \tag{11.2}$$

$$\Delta I = k \times (I - I_{0)} \tag{11.3}$$

The Weber-Fechner Law

Later (1860), Gustav T. Fechner adapted and extended *Weber's* law, encouraged by experimental results. He found that the fraction value should not be a constant, but a logarithmic function [105, p. 290] – as indicated in equation (11.4).

$$\Delta I = k \times \log \frac{I}{I_0} \tag{11.4}$$

The logarithmic relationship between stimulus and perception means that if a stimulus varies as a geometric progression (i. e. multiplied by a fixed factor), the corresponding perception

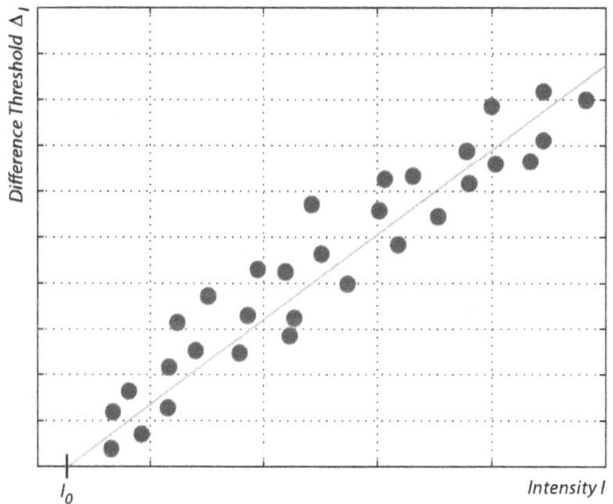

Fig. 11.1: Threshold versus intensity (TVI) plot of Weber's law.

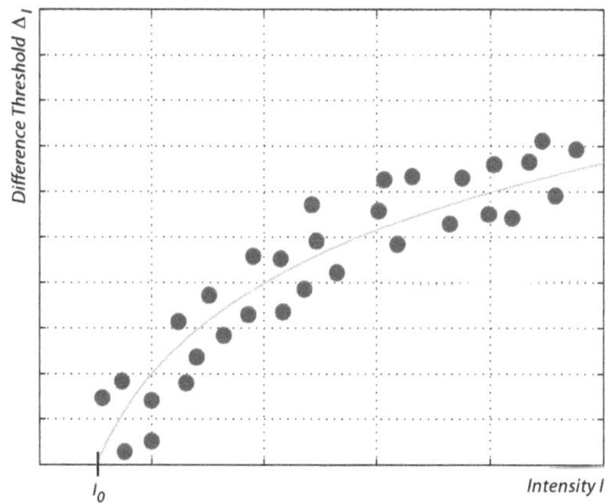

Fig. 11.2: Threshold-intensity plot for the Weber-Fechner law.

is altered in an arithmetic progression (i.e. in additive constant amounts). For example, if a stimulus is tripled in strength (i.e., 3×1), the corresponding perception may be two times as strong as its original value (i.e., $1 + 1$). If the stimulus is again tripled in strength (i.e., $3 \times 3 \times 1$), the corresponding perception will be three times as strong as its original value (i.e., $1 + 1 + 1$). Hence, for multiple increases in stimulus strength, there are only additive increases in the strength of perception.

Steven's Formula

For general application, it has turned out that the *Weber-Fechner* law is often too inflexible [400]; therefore the equation generally used today is *Steven's* formula [398, p. 16]. This formula was worked out in the 1950s based on large-scale experiments with different stimuli and user interviews. Steven found that distinct modalities have completely different curve forms [105, p. 292], e. g. pain has a steeply rising curve (meaning pain rapidly becomes stronger when the stimulus increases), whereas sound or light have only a gently inclined curve form (this means that stronger changes in stimulus are required to be noticed by humans).

$$\log \Delta I = a \times \log(I - I_{0)} + \log k \tag{11.5}$$

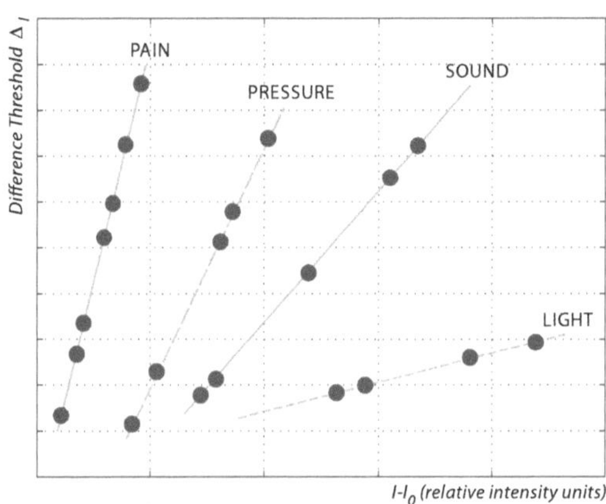

Fig. 11.3: TVI plot of Steven's formula for different sensory modalities.

11.4.3 Stimulus Threshold

Fig. 11.4 shows that there is a strong interrelationship between the stimuli threshold of Pacinian corpuscles and stimulation frequency, with a full range of stimulation varying from about $30Hz$ to $1,000Hz$ [401]. According to Gescheider *et al.* [402], the lowest threshold values (and thus, the highest level of perception) in this U-shaped function of stimulus frequency are in the frequency range of approximately 250 to $300Hz$. The shape of this curve has a strong significance for the definition of appropriate feedback levels in the setup of experimental systems: The perceived stimulation intensity can be forthwith varied not only by adjusting the amplitude level of signals, but also by changing the stimulation frequency.

Contrary to this standard situation are specific applications which should operate with a variation in stimulation frequency and at the same time require a constant level of stimulation. Here, the vibration amplitude had to be readjusted in magnitude based on the actual frequency and considering of the shape of the stimulus threshold graph.

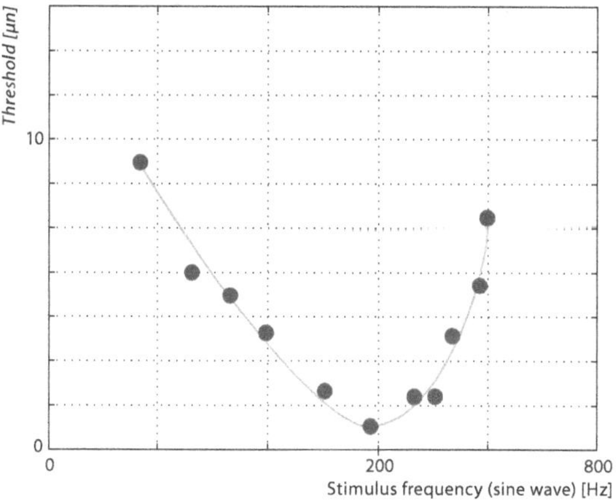

Fig. 11.4: Response behavior of Pacinian mechanoreceptors (PC) in relation to stimuli frequency.

11.4.4 Age and Gender Sensitivity

For a universal Human-Computer Interface based on the sense of touch, it will be required that the stimulation for a certain information item or for a designated level of importance is similar for all qualified persons. Accordingly, age, gender and other factors with potential dependencies must be taken into account.

With age, sensitivity to pressure, vibration, and spatial acuity decreases. On the other hand, response time and threshold rise with increasing age. As far as gender is concerned, and here over virtually every age group, males have faster reaction times than females, and the female disadvantage cannot be reduced by practice[58]. A detailed discussion on age and gender-related influence on the sense of touch and an exploration of related work in this field are presented in Appendix D: "Human Factors" on p. 225.

The influence of the factors age and gender in the automotive domain (individually for the *visual*, *auditory*, and *haptic* modalities) has been investigated in this work, for instance, in the experiment "Age and Gender-Related Studies on Senses of Perception for Human-Vehicle

[58]The male-female difference was accounted for by the lag between the presentation of the stimulus and the beginning of muscle contraction.

Interaction". A summary of the findings is given in the Section "Simulating Real-Driving Performance" (p. 161); the corresponding conference article is [403].

11.4.5 Summary and Impact

Findings from *Weber's* law, *Weber-Fechner* law, and *Steven's* formula are used as a basis for the definition of appropriate and viable parameter settings for the parameters (*i*) intensity, (*ii*) frequency (under consideration of the interrelationship between stimuli threshold and frequency), and (*iii*) activation time, individually for application-specific mappings of vibro-tactile feedback to guarantee accurate perception of stimulation at all times.

Furthermore, the supposed age dependency of reaction-times from vibro-tactile stimulation combined with the fact that the population is increasingly aging need to be taken into consideration in the same way as sex-related deterioration in order to produce a universal solution for touch-based interaction.

12 Experiments

To verify the hypotheses and research questions (as defined in Chapter 5 on p. 30), several experiments have been conducted using custom-built hardware prototypes and software components (a detailed technical description of the hardware systems used for pressure sensing and delivery of vibro-tactile stimulation can be found in Appendix A: "Hardware for Processing Pressure and Vibration" on p. 199).

This research work is focused on the Driver-Vehicle Interaction (DVI) loop as introduced in Fig. 1.1 on p. 5, and in more detail on the investigation of haptic perception and touch-based articulation in the automotive domain. For interaction and communication purposes, system constellations comprising *one vehicle* and *one driver* are considered.

In a vehicle, the seat and backrest of the *driver's seat* are used for implicitly receiving and expressing information. Driver-vehicle input is generated by force sensitive array mats, unobtrusively integrated into the seat. Feedback from the vehicle toward the driver is given with vibro-tactile actuators, again embedded into the driver's seat. It should be noted that the application of vibro-tactile actuators is not limited to a driver's seat and could, for instance, be embedded in control elements, such as the steering wheel, brake and acceleration pedals, throttle control, handbrake, and various other switches. Depending on the (tactile-)stimulated part of the body, different parameters of the vibro-tactile system need to be adjusted (e. g. lower inter-actuator distance or changed pattern for vibro-tactile stimulation to raise attention on the finger tips instead of on bottom or back, lower vibration amplitude for vibro-tactile elements integrated into the steering wheel in contrast to tactors embedded in the seat, etc.).

The studies and tests conducted are presented in the following order: The first part deals with the experiments related to *vibro-tactile input*, which is here the ability of a driver to articulate information toward the vehicle, followed by a description of experiments associated with Vehicle-2-Driver (V2D) output, which covers the opportunity of Driver Assistance Systems to provide information of certain importance to the driving person using one or more sensory channels.

Driver-Vehicle Input

This part of the research activities centered on the information flow from the driver toward the information processing (sub-)systems in the vehicle. Any driver involved in Human-Machine Interaction needs the ability to communicate with that system. With a larger number of appliances and assistance systems in vehicles, and in addition a rising operation complexity of such devices, the demand for distraction-free, implicit articulation, e. g. for identification or authorization tasks or for automatically recognizing a driver's ongoing activities, becomes stronger.

The input channel allows for a differentiation among traditional vehicle control activities, such as steering, operating multimedia devices, route planning or navigating, etc. and a new class of person-related services or activities. For the latter, a further refinement into subclasses is adviseable. This includes (*i*) authorization (permit or deny engine start, configure specific motor settings, activate/deactivate a horse-power regulation service, personal parameters for seat, mirrors, or radio station, personalized car-insurance policy, cause-based CO_2 taxation, fleet management, etc.), (*ii*) security services (secure transactions for mailbox access, mobile banking, online shopping, and concurrently keep passengers from spying personal PIN codes, passwords, etc.), and (*iii*) safe-driving services or "Fitness to Drive" (observe driver's behavior, e. g. fatigue or drunkenness, and initiate actions depending on the driver's momentary condition) is advisable.

Objective

The objective in this cluster of experiments was the development and evaluation of Driver-2-Vehicle (D2V) interfaces, operating implicitly and thus requiring no additional attention of the driver to the task of data transmission. Such an interface could perhaps be used for the automatic adjustment of the settings of a vehicle (attributes like chassis, air conditioning or route guidance system, radio station, but also personalized services such as a car insurance contract or cause-based taxation) in accordance with the personal profile of an identified driver.

The subsections below give an overview of the individual studies carried out in the scope of this work, describe each goal and present achieved results. The findings of the different experiments have been partly published as conference articles (see corresponding references).

12.1 Identification and Authorization

Experiment 1

Driver identification (or authentication[59]) in vehicles is becoming more and more important due to the emergence of personalized services, such as road pricing, personal car-insurance, cause-based taxation, etc. (see Fig. 12.1). Today this identification is done either by traditional approaches (for instance, a pincode or a token in the ignition key) or with biometric identifiers (e. g. retina scan or finger print). The first class of identification systems is not based on any inherent attributes of an individual – and suffers from several drawbacks: Pin codes may be forgotten or guessed by an imposter; tokens may be lost or stolen. On the other hand, biometric identification technologies would allow a higher level of security to be achieved. But systems that are in use today require *active cooperation* of the driving person (a finger has to be put on the fingerprint reader; the eye needs to be aligned to the retina scanner) and therefore create an additional cognitive workload or lead to distraction from the main task of driving (for details see Appendix B: "Biometric Identification" on p. 207).

The aim of this first experiment is a qualitative statement about whether an implicit data acquisition and posture pattern analysis system based on pressure sensor arrays, integrated into the seat and backrest of the driver seat, is viable for use in identification or authorization tasks in both stopped and moving vehicles, or not. The experimental setup followed the suggestions of Jain *et al.* [404, p. 92]. They recommended evaluation of the four characteristics *universality*, *uniqueness*, *permanency*, and *collectability* in order to make assertions on the quality and accuracy of a biometric system (in the evaluation part below, each of these attributes will be analyzed individually).

12.1.1 Experimental Design

For data acquisition two force sensor arrays (FSA) integrated into a car's seat, and linke to a common notebook computer (via USB-interface), were used. The implemented system is universal and could thus be used in any type of car (utility-driven car, sports car with body-contoured seats, comfort station wagon, etc.; see Fig. 12.2) for arbitrary styles of sitting.

To assess the quality and accuracy of the prototype, studies with 34 participants were carried out inside a specific car (the comfort station wagon shown in the far right column in Fig. 12.2), with pressure mats attached to both seat and backrest. The voluntary test persons (students, friends, and department staff) were recruited on the university campus. Most of them were male (79.41%) and 80% ($P_{90} - P_{10}$) of all test persons were between 20 and 31 years old. Male

[59] *Authentication* provides a way of identifying a user. Following authentication, a user must gain *authorization* for doing certain tasks – the authorization process determines e. g. whether the user has the authority to issue a certain command or not (for details see Appendix B starting on p. 207).

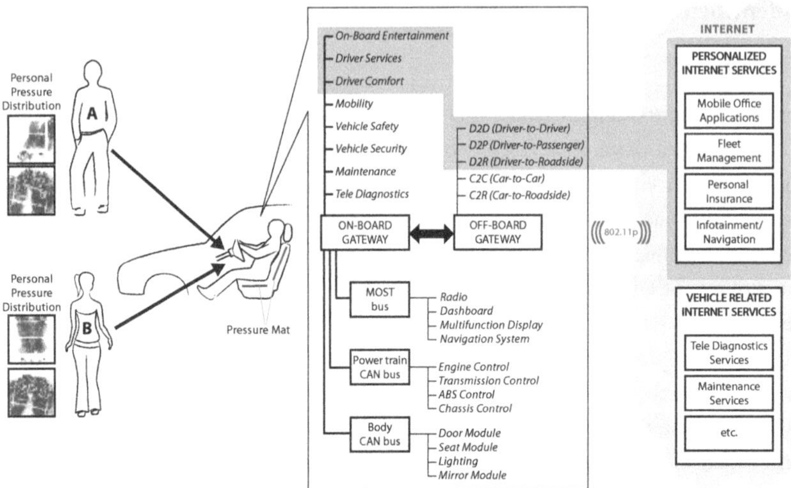

Fig. 12.1: Driver identification from sitting postures and its utilization potential in personalized vehicular services.

subjects varied in weight from 58kg to 104kg, and in size from 160cm to 194cm. The female test persons varied from 46*kg* to 70*kg* in weight, and from 160*cm* to 173*c*m in size (for detailed statistics on test persons, see Riener *et al.* [405]).

Fig. 12.2: The prototype for driver identification, installed in a utility car, a sports car and a comfort station wagon (from left to right).

For initial evaluations, Euclidian distance metrics – a well-researched approach in the field of pattern recognition – were used to match a person's current sitting posture against earlier stored patterns. The experiment was conducted in two stages, (*i*) recording of the training set, and (*ii*) system evaluation with a testing set.

To eliminate the impact of sudden movements during data acquisition in the vehicle, the median of each of the 1,024 sensor values over a series of measurements was calculated[60]. Initial tests showed that using five measurements is sufficient to create a stable matrix of pressure values. To ensure data integrity (according to the Hamming distance) each experiment was repeated four times – resulting in four data sets for each test participant in the database.

12.1.2 Feature Evaluation

The calculation of the *feature vector* which could also be referred to as the *personal sitting profile* was done as a weighted combination of several individual parameters extracted from the sitting posture image of a specific person (as illustrated in Fig. 12.3; the numbers in the circles correspond to the numeration of items in the list below).

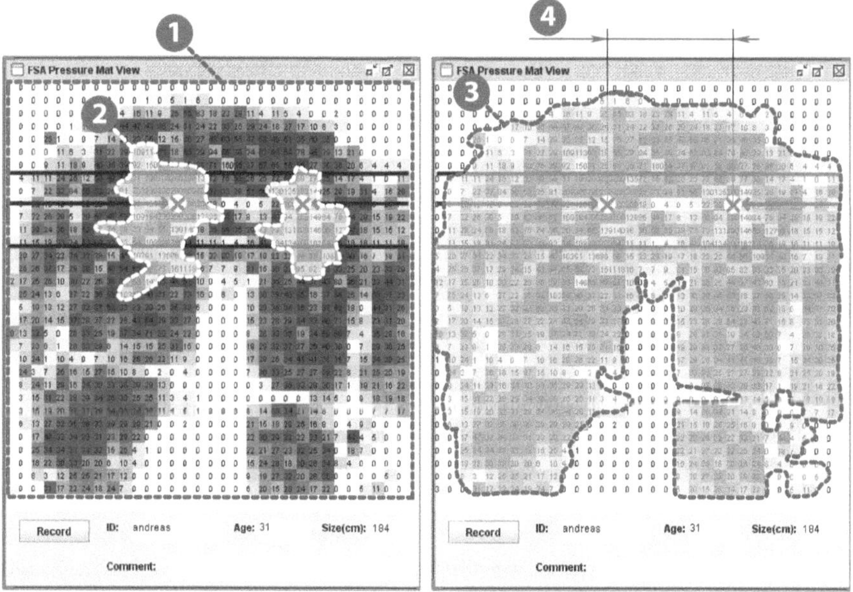

Fig. 12.3: Illustration and localization of parameters used for the feature vector calculation.

The following features were selected and utilized for evaluations regarding person identification from static sitting posture patterns.

[60]The median was preferred against other characteristics, such as the mean, because of its better stability concerning to outliers.

(1) Weight

A weight approximation for the driving person was calculated as an accumulation of the pressure values recorded by the entire sensing area of the mat on the seat (sum of 1,024 equally calibrated sensors). Mostly due to unbalanced load sharing and dead space between the sensors, the total weight could not be exactly estimated from the charged sensors but was a rather vague estimation. Particularly in the area of the thighs on the seat, the pressure value exceeded the calibration range maximum – in this case a correct assessment of the total pressure on the mat was no longer possible. However, sensor mats with higher pressure resolution and lower intersensor distance would probably solve this precision problem.

Prior to the experiment a *weight reference* had to be determined. One person with a given weight was placed in the seat and a large number of consecutive pressure readings were recorded. The mean over all accumulated pressure values (100), factorized with the weight of the person, was used as a reference for further weight estimations (see equations 12.1 – 12.3).

$$\overline{pr}_{NORM} = \frac{1}{100} \sum_{j=1}^{100} \sum_{i=1}^{1024} pr_i = 176.589 \tag{12.1}$$

$$pr_{FACT} = \frac{weight_{real}}{\overline{pr}_{NORM}} = \frac{74.80kg}{176.589} = 0.4235 \tag{12.2}$$

$$W = pr_{FACT} \times \sum_{i=1}^{1,024} pr_i \tag{12.3}$$

(2) High Pressure Area

This attribute considered only high pressure values on the mat and was determined by the number of all sensors exceeding 90% of the local pressure maximum (for the corresponding calibration process see equations 12.4 – 12.6).

$$pr_{min} = min(pr_1, pr_2, \ldots, pr_{1,024}) \tag{12.4}$$

$$pr_{max} = max(pr_1, pr_2, \ldots, pr_{1,024}) \tag{12.5}$$

$$A_{HP} = \sum_{i=1}^{1,024} cnt(pr_i \geq (0.90 \times (pr_{max} - pr_{min}) + pr_{min})) \tag{12.6}$$

(3) Mid to High Pressure Area

This parameter was related to the "High Pressure Area" as described above but counted the number of sensors whose values exceed 10% of the normalized pressure minimum for the actual analyzed data set (equation 12.7).

$$A_{MHP} = \sum_{i=1}^{1,024} cnt(pr_i \geq (0.10 \times (pr_{max} - pr_{min}) + pr_{min})) \tag{12.7}$$

(4) Pelvic Bones

This factor was more sophisticated and included location as well as distance of the thighs on the seat mat. It is rather simple to identify the pelvic bone regions because they produce two areas of very high pressure on the mat. The Euclidian distance between the midpoints (x and y coordinates) of left and right pelvic bone was calculated and used as fourth parameter for the feature vector. The usage of this feature benefits from the following stability characteristics.

(i) **Durability:** The distance between the pelvic bones does not change if people gain weight or wear different clothes (but size and shape does). Thus, this feature is *permanent*.

(ii) **Gender dependency:** It is established that the pelvic bones of males and females are different [406]. The feasibility of discrimination has been shown e. g. by Giles [407] or Stewart [408] and it would gain additional benefit for any person-related application if there is a chance to palpably determine the gender of any individual by a single feature.

12.1.3 Biometric Identification

The two biometric identification features *universality* and *collectability* are supported by the proposed experimental system (as verified in [405]); the remaining two characteristics *permanency* and *uniqueness* need to be investigated in more depth. Primarily, the following preconditions apply to the prototype:

(i) Universality and (ii) Collectability

All of the four evaluated features are based on posture patterns collected from a car seat (two sensor array mats, one on the seat and another one on the back). Any vehicle driver has to sit on this seat while driving; therefore, both *universality* and *collectability* should be guaranteed.

(iii) Permanency

Different clothes, such as beach wear, jeans, or a ski suit, trouser buttons pressing on the mat, or objects like a wallet, a cell phone, or a bunch of keys in the back pocket, have more or

Data Sets	Database	Min x_{min}	Max x_{max}	Mean $\overline{x}$	Median $\widetilde{x}$	Std.Dev. σ	(P_{25})	(P_{75})
Seat mat (normalized)								
34	Normal	24.78	100.00*)	54.79	52.93	12.39	46.45	62.08
8	Artefacts	28.82	58.55	41.89	40.79	8.13	35.59	48.75
Back mat (normalized)								
34	Normal	15.16	100.00*)	41.10	38.43	12.00	32.21	45.56
8	Artefacts	11.61	28.27	20.02	19.47	4.16	17.37	2.45
*) *Maximum difference of any two normal data sets has been set to 100%.*								

Table 12.1: Statistics on experimental data for the test on permanency.

less influence on the accuracy of the identification process. To identify problems regarding the feature *permanency* a special experiment with different artefacts in the back pockets of test participants was conducted (see Table 12.1). Tables 12.2 and 12.3 show the confusion matrices for both seat and back mat for that experiment. The sitting posture variants[61] of a specific subject with different objects in his/her back pockets[62] were compared to each other. The upper triangular part of a confusion matrix shows a distinction between postures with different artefacts. The *scale factor* as defined in equation 12.8 was determined from the maximum difference of any two *normal sitting postures* of the 34 subjects in the database (these are the postures of test candidates with no artefacts in their pockets) and was used as a reference (= 100%) independently for seat and backrest.

$$|max(pr_1, pr_2, \ldots, pr_{34}) - min(pr_1, pr_2, \ldots, pr_{34})| = 100\% \qquad (12.8)$$

Interpretation

The variance calculated from the posture patterns on the back of the driver's seat, with a maximum value of 28.27%, is quite low (see Table 12.3) compared to the variance on the seat. This is because all the artefacts were placed in the back pockets of the test subjects and not directly on the back. On the other hand, the maximum difference between two pressure patterns collected on the seat mat for a person in *normal posture* and the same person with applied *artefacts* in their back pockets is 58.88% – this would be too high for a reliable person identification system

[61] Abbreviations of artefacts in the back pockets: N ... Normal posture, KL ... Bunch of keys, left pocket, KR ... Bunch of keys, right pocket, CL ... Cell phone, left pocket, CR ... Cell phone, right pocket, DL ... Digital camera, left pocket, DR ... Digital camera, right pocket, and DRKL ... Digital camera, right pocket and bunch of keys, left pocket.

[62] All items removed from the pockets of the test participants during the recording of the training set have been included – as a representative subset – in the list of artefacts to evaluate.

	N	KL	KR	CL	CR	DL	DR	DRKL
N	-	29.4	39.8	50.0	46.5	55.6	52.7	58.9
KL		-	38.3	42.6	43.0	48.4	49.9	51.0
KR			-	33.8	33.5	49.0	36.0	47.7
CL				-	34.0	37.0	37.6	45.4
CR					-	36.9	28.8	39.5
DL						-	41.8	34.3
DR							-	31.5
DRKL								-

Table 12.2: The normalized confusion matrix of postures with artefacts for the seat (values in percent).

(it should be remembered that the maximum difference between postures from any two persons is 100.00%).

The feature *permanency* cannot be guaranteed in the evaluated prototype if allowing any type of artefacts in the driver's pockets or even any clothes. The statistical characteristics of artefact-afflicted data sets ($min_x = 28.82\%, \overline{x} = 41.89\%, max_x = 58.88\%$) are similar to those of normal data sets ($min_x = 24.78\%$, $\overline{x} = 54.79\%$, $max_x = 100.00\%$). A comparison of results can be deduced from Table 12.1.

	N	KL	KR	CL	CR	DL	DR	DRKL
N		19.3	20.0	23.7	28.1	23.3	26.3	24.7
KL		-	22.2	20.0	28.4	18.6	26.5	20.2
KR			-	14.8	18.2	17.9	17.0	17.4
CL				-	15.9	11.6	18.9	15.0
CR					-	19.7	18.7	17.3
DL						-	21.7	15.5
DR							-	19.8
DRKL								-

Table 12.3: The normalized confusion matrix for the back mat (values in percent).

(iv) Uniqueness

For determining the *uniqueness* of posture patterns, the feature vector was evaluated with a large number of readings for two specific persons (31 and 105 samples). The results for the attribute *weight* according to the attribute *accuracy* are shown in Table 12.4 (upper rows exhibit raw weight data, lower rows indicate normalized weights after factorization with the reference weight). The mean values give a rather precise estimation of people's weights, but with a large deviation (for instance, of $\pm 5.94kg$ for the test case with 34 readings).

For an evaluation of the feature *uniqueness*, the 34 (105) collected sitting postures were compared to all posture patterns in the database (training set), and a list of deviations was calculated and stored in separate tables. The *rank*, which is an indicator for the accuracy, was calculated according to the guidelines presented in Fig. 12.4 (a rank of 1 means best or exact match of the current feature vector and the value in the database; a rank of 34 means worst match or largest difference – for the first case with 34 data sets contained in the database).

Interpretation

Fig. 12.5 shows the findings of two experiments on *uniqueness*, one series with 31 (dark grey) and another one with 105 (light grey) consecutive readings.

The darker bars in the diagram indicate that the matching of all individual data sets out of the testing set corresponds to ranks 1 to 6. In the worst case, which occured in only 3.33% of the tests, the matching between any posture from the testing set against all (34) postures from the training set resulted in position 6. This demonstrates rather high stability for the sitting posture of that person, but contrary to the overall result the diagram also shows that an exact match (=rank 1) was obtained in less than 25% of the subjects studied (with an optimal value of 100%).

Ranks of Congruence Classes	
RANK LEVEL	DESCRIPTION (RESULT)
1	is the exact match between the current posture pattern of a person and its database-stored counterpart
1 < a < b	a rank of value *'a'* means that the matching between the actual feature vector and the database yields a *'a-1'* better scored but wrong data set (the correct data set was exactly listed on position *'a'*)
b	(with *'b'* subjects in the database) is the worst possible case; the actual reading and its corresponding counterpart got the poorest possible agreement (all other data sets got higher ratings)

Fig. 12.4: Instructions for calculating the ranks (congruence classes).

The light gray bars show the results for the second, larger experiment with 105 readings and are somehow divergent to the first one. An exact match between the testing set and the training set is here given in 48.08%, but the remaining 51.92% are nearly evenly distributed to all ranks

Data Sets	Actual Weight	Min x_{min}	Max x_{max}	Mean $\overline{x}$	Median $\widetilde{x}$	Std.Dev. σ	(P_{10})	(P_{90})
Weight								
105	75.5	137.976	196.339	176.589	177.599	14.530	157.749	193.004
31	80.0	162.937	212.093	192.251	196.602	13.933	165.606	207.200
Weight (Normalized)								
105	75.5	58.797	83.667	75.251	75.682	6.192	67.223	82.246
31	80.0	69.433	90.381	81.925	83.779	5.937	70.571	88.295

Table 12.4: Accuracy of consecutive measurements of two subjects.

from 2 to 34 (86.54% are assigned to ranks 1 to 10, the remaining 13.46% are fragmented into the ranks 11 to 30, with the absolut worst case – rank 30 – occuring in 0.96%).

The evaluation of the studies on *uniqueness* supports the assumption that the identification rate increases with a larger number of identification attempts (experimental results reported an exact match for 31 readings in 22.33% of the cases, and for 105 readings in 48.08% of them, and showed furthermore that the outlier tends to zero).

12.1.4 Results

An *implicit* driver identification method based on the biometrics of sitting was proposed, inspecting sitting posture patterns acquired from pressure sensing arrays on the driver's seat and backrest. Posture recognition, unlike vision detection techniques commonly used in dynamic environments, does not suffer from environmental conditions like brightness or weather. The measuring system is invisible and unobtrusively integrated into the vehicle seat and requires no attentive participation of the driving person. The prototype was evaluated against established features of biometric identification systems – the conclusions for the two non-trivial parameters *permanency* and *uniqueness* are as follows.

Permanency

This feature gives evidence of system stability regarding *noise* caused by the driver, as for instance (*i*) different clothes, (*ii*) artefacts in the back pockets or trouser buttons, and (*iii*) individual characteristics, such as age and weight. The conducted experiments confirmed the assumption that the system is highly vulnerable to objects in the driver's pockets and different clothes.

As it has been found that the features extracted from the pelvic bone regions remain relatively stable compared to other parameters, an experiment incorporating mainly characteristics of the pelvic bones (e. g. *size* and *shape*) should be conducted.

Uniqueness

The results of the *uniqueness* test for two specific drivers were not convincing. In the smaller experiment with 31 posture readings, an exact match among all testing sets and the database-stored training sets was given in only 23.33% of them, but with a low deviation (in any case the posture pattern match leads to rank classes between 1 and 6 with highest values on ranks 1 and 2). A small value for the variance stands for a high system stability (*low noise*).

The second, larger experiment (105 pressure image readings) provided a divergent result. An exact match was given in 48.08% of the cases, the remaining 51.92% are dispersed across almost all ranks with 86.54% inside classes 1 to 10.

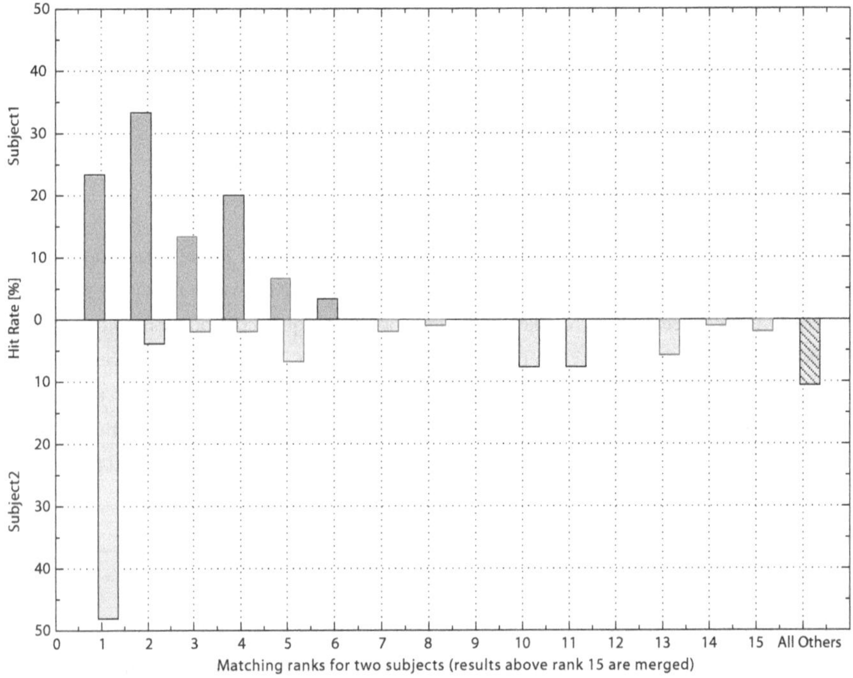

Fig. 12.5: Matching ranks for two subjects (31 and 105 samples).

12.1.5 Improvement Potential

A rework of evaluation functions and/or algorithms based on the results achieved during these experiments should offer potential for major improvements, particularly for the parameter *uniqueness*.

(i) **Feature vector calculation:** The sequence and method of parameter evaluation should be improved, e. g. by (*i*) comparing individual characteristics, (*ii*) applying weight factors, and (*iii*) combining attributes to the final feature vector.

(ii) **High pressure areas:** Excessive loads on the seat mat, particularly for the regions of the pelvic bones, result in corrupted pressure patterns and failures in the current feature vector determination. A more extensive use of low and mid pressure areas instead should compensate for this issue.

(iii) **Filtering:** Applying low-pass filters to captured posture patterns before calculating features should improve the results.

(iv) **Pattern recognition:** The utilization of established pattern recognition techniques or libraries from the face or image detection domain should generate further improvement in the identification rate.

(v) **Attribute combinations:** The quality of any biometric identification system can be improved when combining several techniques, as stated in the section "Biometric Identification" on p. 207. Using additional sensors, such as ECG, GSR, skin moisture, etc., for collecting person-related data should improve the identification system significantly.

12.1.6 General Findings

According to the results achieved and particularly considering improvement potential, the system should be able to uniquely identify and discriminate between persons of a limited group usually sharing a car, e. g. all members of a family, and thus supports **Subhypothesis I.i** defined on p. 34.

However, the performed studies indicate that identification on the basis of sitting posture patterns is not as universal as, for instance, identification based on the eyes (retina scan) or fingers (finger print).

12.2 Activity Recognition

Several constraints need to be fulfilled in order to obtain an accurate and reliable activity recognition system, which is particularly determined by factors such as (*i*) real-time operation, (*ii*) anticipatory behavior (the system must act before the driver), (*iii*) vehicle and environment dynamics (utilizing on-site real sensors and soft sensors[63]), etc.

Compared to the task of driver identification, which is almost entirely based on relatively constant sitting posture patterns, activity recognition is a great challenge. The prototype developed for the first experiment on driver identification can be re-used with marginal adaptations to record dynamic posture patterns including driver and vehicle status information. It is assumed that specific activities in a vehicle (e. g. operating the navigation system, adjusting mirrors, performing steering activities, changing gears, etc.) correlate with selected sitting posture patterns on driver's seat and/or back.

The aim of the "Experiment 2" is an early definition of general conditions and restrictions for subsequent, more complex experiments on activity recognition (e. g. "Experiment 3", p. 138).

Experiment 2

After static posture pattern analyses for the purpose of *driver identification*, the next logical step towards a fully autonomic, implicitly operating Driver Assistance System would be, the use of haptics for *vehicle control* and *steering activities*. Self-adapting vehicles based on the sitting behavior of the driver are accredited with having a crucial impact on (*i*) safety in road traffic, (*ii*) steering a vehicle or operating its control systems, and (*iii*) driving comfort. Such a system would be of particular interest due to the fact that *distraction* from the primary task of driving is continuing to increase.

Before carrying out on-the-road studies to investigate the supposed interrelationship between a driver's activities and his/her sitting postures, activities which would particularly profit from monitoring have to be identified and systematically classified. Along with this key task, applications for dynamic sitting posture capturing need to be developed, and first tests with an early prototype (in a safe vicinity) have to be conducted in order to identify drawbacks and limitations for the final recognition system.

Aside from a feasibility evaluation, the main goal of this experiment was a selection of suitable activities as the basis for a later, specific implementation of an innovative assistance system, automatically initiating activities based on sitting postures, retrieved from the driver's seat.

[63]"Soft sensors" are virtual sensors whose values are delivered, for instance, via the Internet, such as a weather forecast or traffic jam warnings.

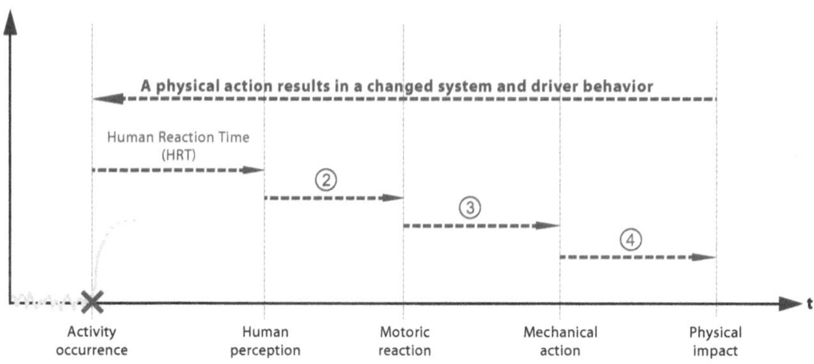

Fig. 12.6: The "Human Perception – Action" loop.

12.2.1 Problem Statement

The state of the art in vehicle control is explicit input which requires active cooperation of the driving person, not only for the main task of driving, but also for secondary or tertiary tasks, and as a consequence may increase the driver's cognitive load or generate (additional) distraction.

Another issue influencing driving performance and/or steering precision arises from the human perception – action loop as shown in Fig. 12.6. It takes some time from the occurrence of an activity to an observable physical impact, either in the vehicle or the environment (there are four stages, each requiring a more or less short period of time for its detection and execution, depending on several constraints, such as the workload of the driver, or the kind and/or importance of the activity, etc.). An implicit data input system in the driver seat, not reliant on the driver's attention and hence, continuously in operation while the driver is seated, has been used to enhance the first stage "Human Reaction Time (HRT)" in this loop by reducing the required reaction time. Therefore, a passive sitting posture recognition system, dynamically evaluating patterns from pressure sensor arrays, has been utilized.

12.2.2 Activity Recognition

The driver-vehicle relationship is not only specified by *basic activities* (often grouped into *classes of activities*, see Fig. 12.7) performed by the driver, but potentially with more significance by *activity chains* or *driving (activity) ontologies*, the latter particularly when investigating cooperative systems (e. g. car-to-car communication).

In-Car Activity Classes					
ACTIVITY CLASS	ACTIVITY	SITTING POSTURE	BACK POSTURE	ADDITIONAL SENSORS	SENSORS (DATA CAPTURING)
DRIVING ACTIONS	Vehicle steering	Normal	Normal	-	-
	Brake	Normal	Lower pressure	OBD	Braking pressure
	Sliding/ABS/ESP	Turbulent	Turbulent	OBD	ABS/ESP warning
	Acceleration	Normal	Increased pressure	OBD	Fuel consumption, throttle valve position
	Slow left turn	Raised pressure L	Raised pressure L	Car-Gateway	Steering wheel pos.
	Fast left turn	Raised pressure L	Raised pressure L	Car-Gateway	Steering wheel pos.
	Slow right turn	Raised pressure R	Raised pressure R	Car-Gateway	Steering wheel pos.
	Fast right turn	Raised pressure R	Raised pressure R	Car-Gateway	Steering wheel pos.
APPLIANCE HANDLING	Blink left/right	Normal	Pressure artefacts left/right	Car-Gateway	-
	Operate the horn	Normal	Short-time increased pressure	ECG	Aggression detection
	Turn high or low beam on or off	Normal	Pressure artefacts	Car-Gateway	-
	Entertainment system (Stereo/MP3/DVD,...)	Pressure artefacts	Pressure artefacts	-	Microphone
	Adjust speed control	Pressure artefacts	Pressure artefacts	Car-Gateway	-
	Navigation system	Pressure artefacts	Pressure artefacts	-	-
PERSONAL (VEHICLE UNSPECIFIC)	Conversation	Dynamic change	Dynamic change	Camera	Face tracking
	Cell phone call	Normal	Raised pressure in one shoulder	Camera	Arm tracking
	Look into road map	Normal	Lower pressure	Camera	Face tracking
	Eating/Smoking	Normal	Normal	ECG, Smoke detector	Raised HF
	Grooming	Normal	Normal	ECG, Smoke det.	Raised HF

Fig. 12.7: Activities, activity classes and corresponding sitting postures for an application in the vehicular domain.

The focus in this research work is on *activities*. However, the main elaboration will be preceded by a short overview of related work on *ontologies* and *activity chains* in the context of this work.

Driving Ontologies

Schlenoff and Gruninger [409] investigated an on-road driving ontology based on an extensive study of driver task descriptions. For each of the behaviors in the ontology, a vocabulary of task commands and a hierarchy of behavior generation processes were developed and evaluated.

Qi *et al.* [410] presented a Mobile Information Service Platform (MISP) based on a Multidomain and Multilingual Ontology (MMO) and associated Query Language (MMQL) with the ability to provide domain-specific semantic and linguistic knowledge. Several experiments (e. g. on a traffic congestion information service or a driving route planning service) showed that the MISP can provide valuable services to (mobile) users. Moreover, the researchers reported that the described services are already in use (in China).

Fuchs *et al.* [411] presented an OWL[64]-based context-model for abstract scene representation of driving scenarios (particularly cooperative Driver Assistance Systems (DAS) demand a common domain understanding of scene representation to enable information exchange between vehicles), considering the operation environment, the driver, the car with its assistance systems, and (national) traffic regulations. The final aim for the model is to serve as a common basis for domain understanding, information sharing and decision making (rule definition and mapping are still in progress).

Baumgartner *et al.* [412] proposed an approach to combine concepts of situation and/or context awareness in road traffic management in an appropriate ontology and furthermore demonstrated its applicability in a number of real-world traffic management scenarios.

Georgeon [413] analyzed traces of activities (sequences of events describing the interaction of a driver with his/her environment) for modeling cognitive schemes of operators based on an ontology of "symbols".

Activity Chains

Activity chains [414] in the domain of driving have up to now only been investigated for traveling or commuting in general (Stopher *et al.* [415], Axhausen *et al.* [416, p. 96], Maerivoet *et al.* [417]) e. g. driving routes such as (*i*) home, (*ii*) work, (*iii*) shopping, (*iv*) home, (*v*) leisure, (*vi*) home, etc. [418, p. 7]. However, it could be imagined that the task of driving itself can in some respects be described by activity chains (e. g. (*i*) look into the rear mirror, (*ii*) apply the brakes, (*iii*) set the direction indicator, (*iv*) turn off, or (*i*) shift a gear down, (*ii*) look into the rear mirror, (*iii*) set the left direction indicator, (*iv*) accelerate, (*v*) overtake, (*vi*) set the right direction indicator, etc.).

The aspect of using *activity chains* for describing driving activities is novel and has to be investigated in detail beyond this research work.

Activities and Activity Classes

A system automatically reacting to a driver's continuous activities requires a common database (the "training set") for the detection of situations and a selection of appropriate actions. The reference database was built up from activity-posture pairs (each analyzed activity is associated with a corresponding sitting posture pattern).

An enumeration of all feasible vehicle-handling activities would result in a long list of items (such as, for instance, partially given in [72], [6], [166], [169], [315], [419], or [420, p.4]). A shorter list of activities, considering only a limited number of items collected through a brainstorming session (see Fig. 12.7), was used in this experiment. After in-depth refinements,

[64] W3C Semantic Web, Ontology Web Language, http://www.w3.org/2004/OWL, last retrieved August 30, 2009

the final version of this list – used as basis for this research work – was prepared and is presented as Fig. 5.2 on p. 33.

It should be noted that the different activity classes are not self-contained – a crosslink between domains is clearly discoverable (considering, for instance, the two classes "Navigation system" and "Entertainment" in Fig. 5.2).

Driving Activities

The first detailed employment of activity recognition focused on the primary task of driving. The following eight activities were identified as the most crucial ones for the class "driving actions" (see Fig. 12.7): (*i*) ordinary cruising, (*ii*) standard braking situation, (*iii*) braking with the help of assistance systems (ABS, etc.), (*iv*) vehicle acceleration, (*v*) to (*viii*) turns with low and high speed in both directions left and right.

Fig. 12.8: The image on the left-hand side indicates a *soft* turn, the right picture shows a *hard* turn (both snapshots show a video of the test run together with the synchronized sitting posture).

For an initial investigation regarding the potential of such an implicit Driver-2-Vehicle (D2V) feedback system, a prototype integrated into a sports car was built up, utilizing (*i*) force sensor array mats on the seat and back, (*ii*) a video camera inside the car capturing the driver's field of vision (through the windshield toward the road), (*iii*) a GPS device to track the trace of the journey, and (*iv*) an OBD interface recording vehicle specific data such as speed, engine RPM, temperatures, etc.

Acceleration Forces

When steering a vehicle, different forces of acceleration – with interdependencies – are present. The x-axis of the car corresponds to the direction of driving and points in the same direction as the acceleration force acc_x, the y-axis represents the lateral force with positive values acc_y pointing "toward the center of the turn", orthogonally to the x-axis (=centripetal force), and the centrifugal force $acc_{y'}$ points in the opposite direction as acc_y (has an equal value, but the opposite sign).

A series of test runs were conducted in public road traffic during regular hours. The inspection of recorded dynamic sitting posture traces, synchronized with a video showing the driving journey from the view of the driver, showed some correlations between sitting postures and driving scenes. For instance, for cornering driving situations it seems that there are two classes of posture-activity mappings, separated by a certain speed of driving.

In low-speed driving situations, it can be observed that the lateral acceleration force of the vehicle affects the sitting posture of the driver (a deflecting force – the centrifugal force – moves the driver to the opposite side of the turn which is discoverable from the sitting posture). The driver him/herself is not actively aware of this and thus, does not try to compensate for this force. In faste turns (beyond a personal barrier limit, “break-even speed” v_{BE}) it can be realized that the driver becomes aware of the (much stronger) centrifugal force just before it actually takes effect and tries to balance it (anticipative posture).

The subject of investigation was validating if the shift in the sitting posture before high-speed cornering could be universally used in a vehicle assistance systems, for instance, to warn the driver if he/she was about to miss a curve or to assist the activity of steering.

Another question to be dealt with in this context was if the the driver is more concentraged in situations driven with speeds higher than the break-even speed, compared to sections driven at lower speed.

In defining a universal experiment, incorporating all the driving activities listed in Fig. 5.2 should allow further correlation between steering and body postures to be identified. For this reason, a specific course on a public road was selected as shown in Fig. 12.9, and fixed segments for the individual driving activities acceleration, braking, turns, etc. were defined.

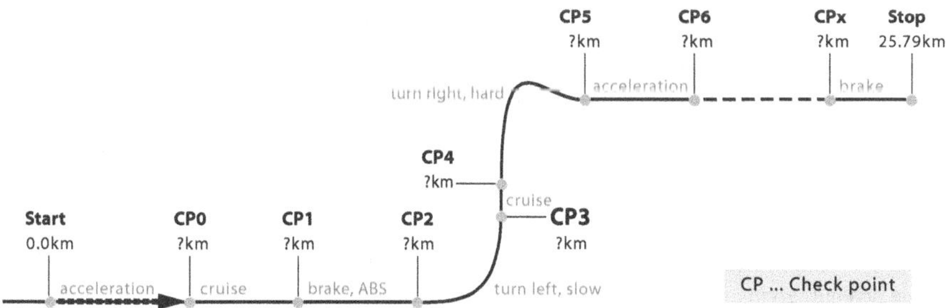

Fig. 12.9: On-the-road experiment for identifying correlations between body postures and steering activities.

To verify all of these questions, a professional vehicular prototype was prepared for testing in on-the-road scenarios. Experiment 3: “Intelligent Vehicle Handling: Steering and Body

Postures While Cornering" dealt with the investigation of cornering situations, driven on a well-defined race course with precise system specifications and preconditions.

12.2.3 General Findings

The early results of activity recognition based on drivers' sitting postures, as presented in experiment 2, do not allow for qualitative assertions of the potential of such enabled Driver Assistance Systems. Contrary to the initial deliberation that most driving activities do not depend on the driver, first evaluations of on-the-road tests showed that almost any activity is reliant on the driver's behavior.

Subhypotheses I.ii and **I.iii** are both not directly supported by this experiment. However, the previous findings prioritize investigation into a person-dependent monitoring of driving situations (e. g. to identify pairs of a personal sitting posture and a related driving activity with good prospects for universal, reliable utilization) in more detail.

Experiment 3

This study presents the results of an experiment founded on the preliminary findings of the preceding experiment 2 ("Driver Activity Recognition from Sitting Postures"). As a consequence of its complexity, the proposed setup of the experiment (definition of the race course as shown in Fig. 12.9 and activities to inspect as listed in Fig. 5.2) was changed to focus only on variants of turn operations, or more specifically considering only left turns in acceleration and brake sections captured on an orbital counter-clockwise race course (advantages: almost only left curves, same course for all test participants, optimized boarding and take-off times because they were at the same place and required no ride back after an experiment).

Vehicle handling and control are an essential aspect of intelligent DAS and have initiated the engineering of the upcoming generation of "smart cars". The significance of this topic has, for instance, been addressed by Ungerer[65], who characterized "smart mobility" or Car-2-Car (C2C) and Car-2-Roadside (C2R) interaction as one of the three main application challenges for technical computing in the next decade. Of particular interest in this field of research are Driver Assistance Systems that help to relieve the driver of manipulative and cognitive load while driving, giving rise to new modalities in Driver-2-Car (D2C) and Car-2-Roadside (C2R) interaction.

Steering, as one of the most crucial manipulative and cognitive issues in car handling, challenges the driver with a complex, multidimensional interaction process. Both driver experience

[65]Theo Ungerer held a keynote on "Grand Challenges of Technical Computing" at the 2008 conference on Architecture of Computing Systems (ARCS'08), http://arcs08.inf.tu-dresden.de/?n=p, last retrieved August 31, 2009.

and the contextual setting (road and traffic conditions, weather, visibility, etc.) determine car handling performance. Hence, in order to investigate vehicle handling and driver engagement, dynamic models have to be developed, capable of real-time driver activity and multi-sensor based context recognition.

The main objective of this experiment, motivated among others by basic research reported by Andreoni *et al.* [244], was a substantiated statement about the "readiness for proactive behavior[66]" of a Driver-2-Vehicle system. The verification was to be carried out with the two kinds of sitting behaviors identified in experiment 2: (i) the driver has low or moderate ambition to compensate for centrifugal force while cornering at low speed, and (ii) a driver's readiness to compensate centrifugal forces is high in cornering situations driven at high speed (see Fig. 12.10). The value for the "break-even speed" v_{BE} is driver dependent and determined by factors like the person's weight and/or body figure.

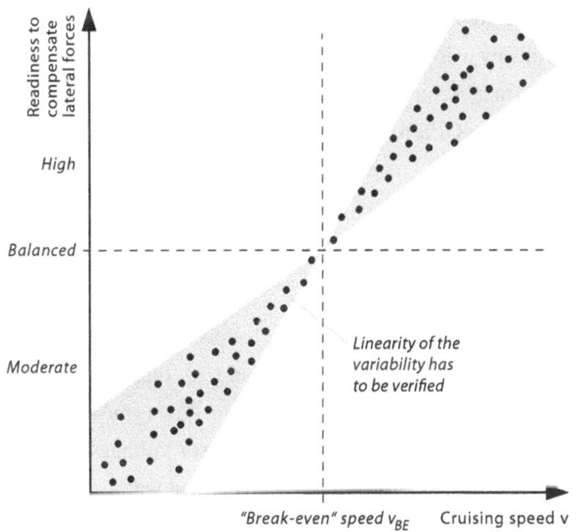

Fig. 12.10: Hypothesis on a correlation between body posture and cruising speed.

In particular, it was investigated how a driver's endeavor to steer precisely interferes with lateral acceleration (centrifugal force) while cornering, based on the following hypotheses:

(i) Cruising speed is correlated to body posture in cornering driving situations as observations from previous driving studies suggest (Fig. 12.10).

[66] Proactive behavior here is the ability of a driver's assistance system to control the vehicle adaptively, thus before explicit activities of the driver actually take place.

(*ii*) Increasing lateral (or centrifugal) force while cornering leads to a higher cognitive load for the driver. The risk of casualties can be reduced if the interrelationship between body postures and cruising speed is taken into consideration.

(*iii*) The time gap between a changed body posture and the appearance of the corresponding activity depends on the level of driving experience and tends to zero for a professional driver (on a familar driving route).

12.2.4 Experimental Design

For data acquisition, the "FSA Dual High Resolution System" from Vistamedical was again utilized (for system details see Appendix A: "Pressure Sensing" on p. 199). Before its initial use, a reference weight had to be applied, using the same method described in experiment 1 (parameter "Weight" on p. 124).

Apart from the values from the pressure-sensitive mats, data from additional sensors had been captured during the trials by using standalone software components with a dedicated timestamp field in their data file format. Prior to the evaluation, data from all individual sensory channels were synchronized, using the GPS time field as common reference.

Sensor Details

The following list provides an enumeration of additional sensors utilized in the evaluation environment.

(*i*) An OBD interface[67] was used to capture the following vehicle-specific data: Cruising speed (km/h), engine speed (rpm) and load (%), throttle position (%), coolant temperature (^{o}C), intake air temperature (^{o}C), air flow (gm/sec.), and a number of motor-specific data such as fuel trim, various sensor voltages, etc. For the experiment described here only the parameter *vehicle speed* was used.

(*ii*) A GPS receiver XGPS BT-929 with SiRF Star III chipset mounted nearby the rear window was used to get precise vehicle positions. The GPS time field was consulted as external synchronization basis.

(*iii*) Acceleration and centrifugal forces were recorded with Inertiacube3[68] high-precision accelerometers placed near each axis of the vehicle.

(*iv*) Videostreams from three Sony DCR-HC96E camcorders were used for a later visual examination of test runs.

[67] On-Board Diagnostics (OBD) refers to a vehicle's self-diagnostic and reporting capability.

[68] Intersense Inertiacube3 for precise 3-DOF orientation tracking, http://www.isense.com/products.aspx?id=44&, last retrieved August 30, 2009.

(*v*) A wireless 3-point ECG device "Heartman 301" from Heartbalance AG[69] was used to record a driver's vital data, such as heart frequency or heart rate variability. Due to measurement errors caused by moving artefacts, ECG data were not used in the evaluation of this experiment.

Pressure Sensing

FSA11, FSA12, FSA13, ..., FSA44 (see Fig. 12.11) indicate regions on the force-sensitve mat for estimating the sitting posture of a driver, and subsequently his/her direction of leaning. Each value represents the sum of all 64 (8*x*8) sensor values in the specified region. A further aggregation of the eight "squares" on the left side of the mat to the region *left* (as indicated by the dashed rectangle in Fig. 12.11), and the eight squares on the right side to the region *right* has been used in this experiment to easily distinguish between the two leaning directions.

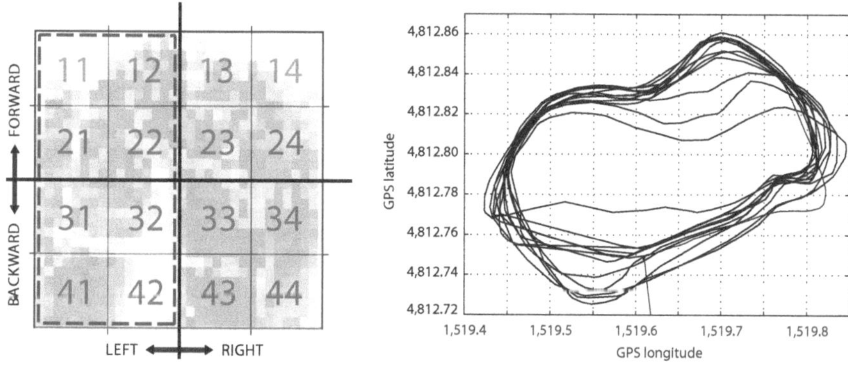

Fig. 12.11: Fragmentation of the pressure-sensitive mat to determine the direction of leaning.

Fig. 12.12: The GPS trace of one specific test run on the Wachauring Krems, Austria

In order to get representative values for the "direction of leaning", namely to compensate for personal sitting behaviors, the disjunction between the regions *left* and *right* should ideally be not in the middle of the mat, but at the midpoint between the pelvic bones. However, previous tests showed that there is almost no difference between the absolute middle on the mat and the midpoint calculated from the pelvic bones and therefore the more easily calculable absolute middle has been used as the border limit for the two regions.

Dynamically evaluated sitting postures on the seat mat were used to calculate the present direction and degree of leaning, where *leaning* was defined as normalized values in the range

[69] Heartbalance AG, http://www.heartbalance.com/, last retrieved August 30, 2009.

$[-50\%, 50\%]$ with a value of zero representing sitting in an up-right position. *Lean left* was defined as a deviation from the initial sitting position (or corresponding posture pattern) to the left side. Analogously, *lean right* was defined as a deviation of the sitting posture pattern toward the right side (it should be noted that *left* and *right* are with respect to the vehicle's direction of motion as defined in the Paragraph "Acceleration Forces" on p. 136). If the (dynamically evaluated) total pressure on the *left* side of the mat is higher than the pressure on the *right* side (see equation 12.9), the driver is inclined to the *left*, and vice versa (equation 12.10 is an indicator for an inclination to the *right* side).

$$pressure(right) - pressure(left) \in [-50, 0[\tag{12.9}$$

$$pressure(right) - pressure(left) \in]0, 50] \tag{12.10}$$

Further information regarding sensor parametrization, implementation of smoothing functions, interpolation mechanisms, etc. is found in the paper of Riener *et al.* [421, pp. 7–9].

Fig. 12.13: Driving studies for the experiment in activity recognition, conducted on the Wachauring Krems, Austria.

12.2.5 Evaluation

To verify the hypotheses on "experienced driving", a field study was conducted in the driving safety center "Wachauring Krems", Austria[70] on a specified race course with a length of about $1,150m$ (see Fig. 12.13 for some pictures regarding the execution of the tests). The experiment was focused on driving situations while cornering left or right in curves with different

[70]http://www.oeamtc.at/netautor/pages/resshp/anwendg/1104290.html, last retrieved August 31, 2009.

radii (unlike to the limitation of only one type of curve being present on the course, a number of different types of turns are distinguishable in *real* road traffic, e. g., right-angled streets or crossings, freeway entrance and exit ramps, U-turns, banked corners or normal curves).

Time	FSA11	FSA12	...	FSA44	Accelerometer1			v	GPS ϕ	GPS λ
ms	kPa	kPa	...	kPa	x	y	z	km/h	GRD	GRD
540,687	5.444	19.144	...	0.000	-2.195	7.427	-3.456	118.648	1,519.736	4,812.780
540,890	5.420	19.026	...	0.004	-1.698	9.657	-0.201	119.528	1,519.739	4,812.781
541,109	10.086	24.365	...	0.004	-1.466	8.422	0.101	120.477	1,519.472	4,812.782
541,640	12.104	28.130	...	0.020	-0.368	8.729	3.824	122.778	1,519.778	4,812.786
541,828	12.081	28.277	...	0.000	-3.925	8.606	-3.213	123.593	1,519.751	4,812.786
...	...	...	...	...	...	...	...	...	...	...

Table 12.5: Tabular specification of features for dynamic activity recognition.

Prior to the the experiment, the test drivers were informed about the purpose of the study and, moreover, received instructions on "how to drive" from the research group's staff.

Data Gathering

Table 12.5 illustrates the tabular structure of all characteristics used in the experiment with corresponding measurement units given in brackets in the subsequent list (the data of the various sensor records were time-aligned and filtered in a pre-processing step to meet the formats of the statistical analysis toolset): (*i*) Time (*ms*), (*ii*) to (*xvii*) FSA11, FSA12, ..., FSA44 (*kPa*), (*xviii*) to (*xx*) accelerometer data in x, y, and z (m/s^2), (*xxi*) cruising speed (km/h), (*xxii*) GPS data latitude ϕ (*GRD*) and (*xxiii*) GPS data longitude λ (*GRD*). Furthermore, the table shows an abstract of five real data sets out of the entire data table of 3,786 rows used for the evaluation.

12.2.6 Discussion

Fig. 12.14 shows a scatter plot of one entire test run. Almost all curves in the counter-clockwise racing circuit were driven left; this explains the majority of data points above the zero line for low-speed driving and means that centrifugal forces were not compensated for by the driver (the driver was inclined to the right, which is opposite to the vehicle's centripetal force directed to the left). The region of points with a strong right inclination in the range 20% to 30% at vehicle speed around zero can be interpreted as resulting from the briefing task before starting the test run. In this stage, the driver was leaning strongly towards right, due to communication with the co-passenger. The dense plotted points in high-speed sections provide no evidence about a possible correlation between lateral force and speed, because (*i*) high speed is usually obtained when driving straight ahead (centrifugal forces are around zero), and (*ii*) the ride was turbulent – inclination of the driver was influenced by curbs, road holes, a bumpy road, etc.

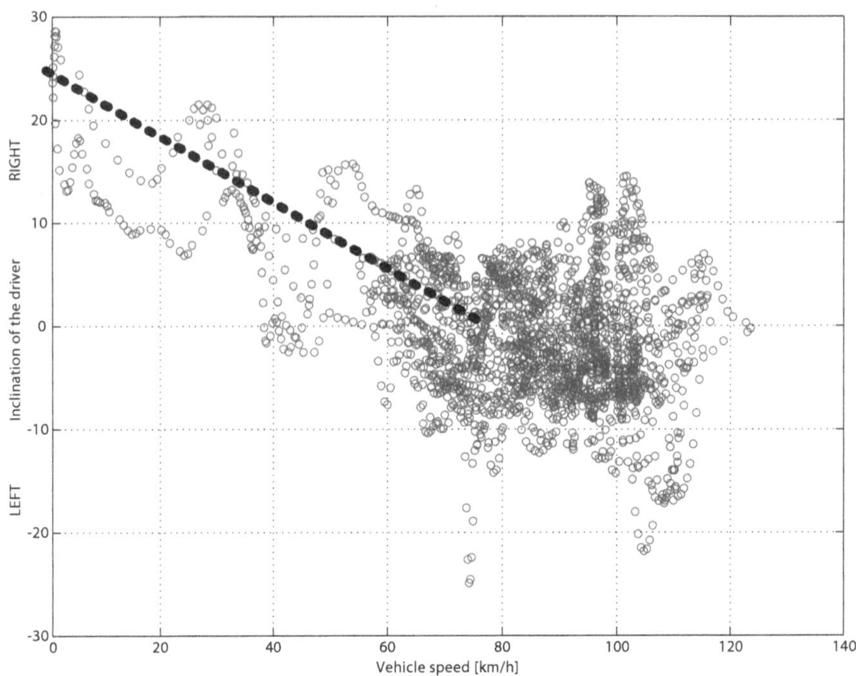

Fig. 12.14: A driver's readiness to compensate for lateral acceleration forces in one test run (several laps).

Compensating for Lateral Acceleration Forces

Fig. 12.15 shows an evaluation of the same test run with filtered data sets. All data points originating from driving scenes with lateral acceleration $acc_y \geq 0$ (straight ahead or right turn sections) and vehicle accleration force $acc_x < 0$ (negative acceleration or braking sequence) have been removed – the remaining database contains only data sets of left turn sections driven with constant or increasing speed.

In this plot it can be recognized that in low-speed driving regions the vehicle's lateral acceleration force is again not compensated for by the driver and therefore he/she is deflected to the right side (it must be remembered that only left turns are considered). But contrary to the first scatter plot (Fig. 12.14), in the current plot an increased readiness of the driving person to compensate for the centrifugal acceleration force can be discovered above a certain driver-specific speed v_{BE} ("break-even" speed). This is represented in Fig. 12.15 by an inclination of the driver in the same direction as the lateral acceleration force of the vehicle. A more representative correlation of lateral acceleration (acc_y) and the corresponding driver's direction of leaning can be recognized in Fig. 12.16.

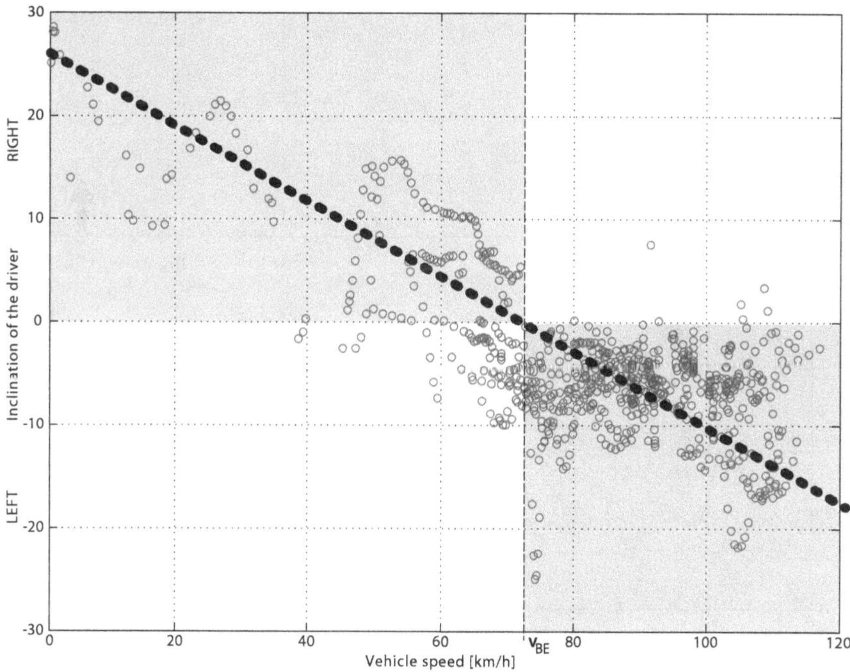

Fig. 12.15: Correlation of body posture to vehicle speed.

The plot of sensor data from the different sources (*i*) OBD (vehicle speed), (*ii*) accelerometers, and (*iii*) pressure sensitive mats (body posture) over the time shows a close agreement between the curves of lateral acceleration and the leaning direction, considering only high-speed driving sections ($\geq v_{BE}$). The figure indicates that the driver is leaned to the left side only in driving sections where the lateral acceleration force of the vehicle is positive ($acc_y > 0$), which means that the lateral acceleration force pushes left, too. This confirms the first supposed hypothesis (for this experiment) on a correlation between cruising speed and body posture (above a driver-specific vehicle speed v_{BE}). For the remaining two hypotheses, formulated in the introduction section of this experiment, no qualitative conclusions can be drawn – they have to be evaluated in additional experiments.

12.2.7 General Findings

This experiment explored a driver's readiness to compensate for lateral forces in cornering driving situations as a function of *lateral acceleration* (accelerometers) and *vehicle speed* (onboard diagnostics interface), using a real-time data acquisition system. The results of this study

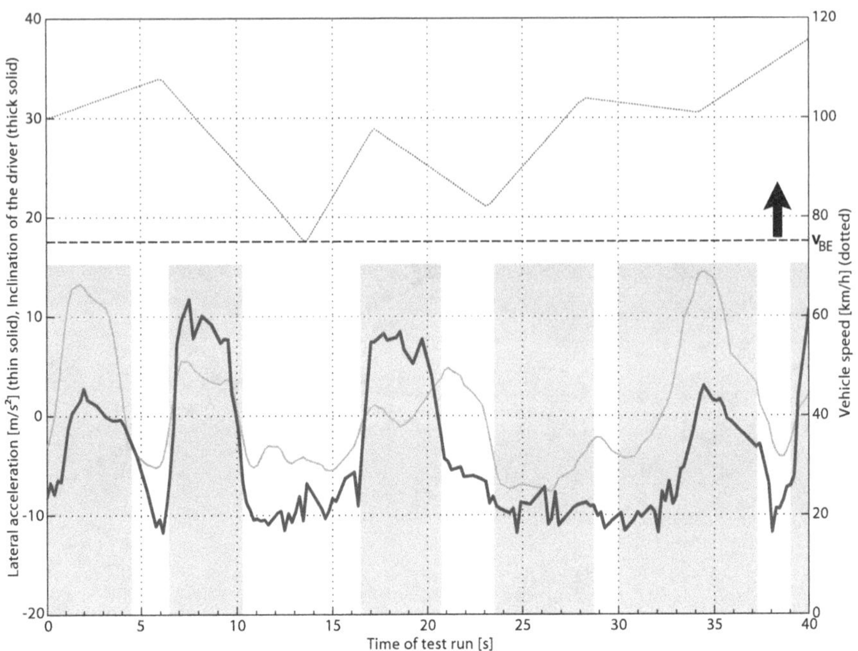

Fig. 12.16: The visual analysis shows that the path of a body posture matches the curve of lateral acceleration.

support **Subhypothesis I.ii** (p. 34) and furthermore encourage investigation into new modalities in Driver-2-Car (D2C) interaction.

The deduction of driving situations before they happen would allow for improvements of Driver Assistance Systems[71] or add-ons for intelligent vehicle control[72]. Furthermore, the experimental results give a first idea of how a system satisfying **Subhypothesis I.iii** ("[..] using posture patterns from the driver seat as an implicit, unique channel for vehicle control") might operate.

[71] Systems capable of enhanced steering precision, allowing reduced or avoided over or under-steering, or offering proactive notification of Electronic Stability Control (ESC) systems, etc.

[72] Adaptive steering angle or chassis frame or a service preventing wheels from spinning (when accelerating and/or cornering).

Vehicle-Driver Feedback: Haptic Perception

Apart from a driver's input capabilities, the second substantial building block to consider in implicit Driver-Vehicle Interaction systems is the perception or output aspect. Almost any assistance system or activity requires the means for user notification. In general, the five sensory modalities (vision, hearing, touch, taste, and smell) are capable of delivering information to the driving person – the first two senses are overused today while the last two make only little contribution to feedback in the vehicular domain. The remaining sensory modality *haptics* or the sense of touch has been utilized as an additional notification channel for Vehicle-2-Driver (V2D) output. Haptics is still underused in vehicles today, but offers potential for novel in-car notification systems.

Objective

This part deals with the application of the sense of touch for assisting the sensory modalities *vision* and *sound*. For this purpose, several experiments were defined with the aim to investigate the applicability of *haptics* compared to the other two modalities (*visual*, *auditory* senses) regarding parameters like response time or interaction performance, probability of distraction, required level of attention, etc.

A second objective was an evaluation of *vision*, *sound*, and *touch*-based interaction with respect to the substantiated fact that interaction using almost any sensory modality is affected by a person's age and gender.

To verify the research hypotheses stated in the Section "Hypotheses and Research Questions" on p. 30, several experiments regarding performance, accuracy, reliability, etc. of vibro-tactile notifications have been conducted – an overview with their individual research goals and the most significant findings is given in the subsequent sections.

12.3 Dynamic Adaptation of Vibro-Tactile Feedback

Experiment 4

Both sensory modalities *vision* and *hearing* are very often highly charged when utilized in the vehicle due to an increasing number as well as a rising complexity of diversified assistance systems – these systems, originally developed for "supporting" the driver in his/her tasks, increasingly lead to operation errors caused by cognitive overload.

The sense of *touch* as an additional channel for information presentation should have the ability to relieve the cognitive load on these two channels by providing feedback in a natural, intuitive, and non-distracting manner [168]. In particular, the sense of touch offers potential for several reasons (Brown, [186]):

(*i*) A rising number of ADAS, combined with an increased operation complexity leads to high cognitive load for the driving person, as stated in subsection "Complex Vehicle Handling" on p. 46.

(*ii*) The visual sense is often affected by factors like reflecting sun, poor visibility or different day and night-vision capabilities, as described in the subsection "Notification-Induced Driver Distraction" on p. 72. A second issue is the accommodation of the eye – changing the focus from far (the road) to near (the dashboard) takes time and particularly older drivers are strongly affected by the task of focusing [303]. Cardullo [181] reported that vision is affected in high-acceleration environments and that the result leads to performance deterioration.

(*iii*) Auditory feedback could easily be missed and is relatively fast considered annoying (e. g. the indicator for fastening the seat belt). Moreover, most driving situations are characterized by ambient, distracting noise originating from (*i*) the engine or environment, (*ii*) communication between passengers, (*iii*) cell phone conversations, or (*iv*) the car stereo (see also the section "State of the Art in In-Car Interaction" starting on p. 41).

(*iv*) Certain feedback options established in traditional computer interaction are not appropriate for utilization in vehicles [43] (further information can be found in subsection "Haptic Interaction in Vehicles" on pp. 58).

Considering the quality characteristics of the sense of touch with respect to the limitations stated above (it is e. g. independent of lighting conditions and not affected by background noise or superposition of voice[73]) legitimates the application of *touch-based interaction* in the

[73]Wilkins [185] stated that high levels of noise may be associated with higher accident rates.

automotive domain in order to assist *vision* and *hearing*. Nevertheless, *haptics* also has its drawbacks – it is, for instance, reliant on a driver's personal attributes (different sizes and weights of drivers would result in a changed haptic perceptivity, which in succession could result in erroneous interpretation of vibro-tactile feedback).

12.3.1 Requirements for a Vibration-Based Seat System

The seat is continually in contact with the driver and is responsible for driving comfort or discomfort and safety. A detailed investigation into vibration sources in vehicles and their impact on the driving person is given in the subsection "The Driver Seat: Suitable for Haptic Notifications?" starting on p. 61. Summarizing this related work study, much research effort has been spent on vehicle seat design during the past years in order to cope with the stated risks of vibrations in the seat. One feasible solution has been presented as the so-called "active seat" which compensates for vibrations originating on the road or in the power train.

The subsection "Haptic Interaction in Vehicles" (p. 58) deals with the fact that it is particularly difficult to interact with small-sized devices in vehicles (e. g. due to a bumpy journey by reason of road condition). Similar interaction is happening in the car when the driver wants to control technical appliances like a navigation system or an on-board computer during the ride. These concerns can be (at least partly) resolved when incorporating touch sensors in the seat, employed for dynamically reconfiguring haptic feedback based on (i) a driver's sitting posture and (ii) vibration forces from the environment.

The Seat System

The prototype used for vibration sensing and transmission in the experiment under review was an extension of common vibro-tactile systems. It was built up from a conventional vehicle seat, supplemented with (i) pressure-sensitive, and (ii) force-feedback array mats, both similar in size and shape, and one placed on top of the other (for an outline see Figure 12.17). The specification of the proposed system includes the requirements that stimulation has to be (i) (much) stronger than the permanent vibrations resulting from the environment, (ii) highly different in its feeling as opposed to underground vibrations, and (iii) fully perceivable by drivers of any size and figure. The specifications (i) and (ii) can be ensured, at least when using an active technology to compensate for the vibrations, requirement (iii) can also be guaranteed because both seat and back mat will be fully covered with tactor elements.

In more detail, this research work focused on an investigation of intuitive perceivable haptic feedback in vehicle seats. Starting with static and dynamic evaluations of sitting habits on a large number ($N = 34$) of test persons (voluntary students, friends, and department staff), a system automatically and dynamically reconfiguring itself with the objective of providing accurate feedback on the *vibro-tactile seat* is presented.

Finally, the application of this system should assure that user notification fulfills the following requirements:

(i) **Independence:** Haptic output is autonomous from a person's driving behavior (sitting position, shape of back or bottom, etc.).

(ii) **Calibration free:** The *vibro-tactile seat* requires no calibration at all, thus is maintenance free and can be immediately used by any driving person.

(iii) **Weight sensitive:** The weight of the driver (calculated from the posture pattern on the mat) is used to control the vibration amplitude to ensure a similar feedback experience for every person.

(iv) **Environmentally aware:** Vibration forces arising from a vehicle's engine, from road conditions (such as road holes), or from lateral forces are compensated for, so as not to affect the driver's haptic perception.

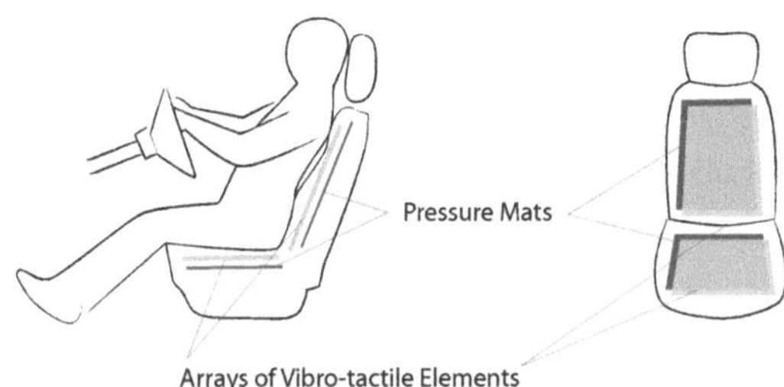

Fig. 12.17: The vibro-tactile car seat consisting of pressure-sensitive mats and arrays of tactile elements on both seat and back.

12.3.2 Touch Sensation

Linjama *et al.* [422] observed that one of the greatest challenges for the application of haptics is to guarantee that haptic cues comply with cues provided by the other sensation channels (mostly vision and hearing). This is caused by the fact that a human's perception uses all the sensory modalities in an integrated manner (Pai, [423, p. 2]). An example applied to the current experiment would be, for instance, the utilization of haptic feedback for indicating turns. The vibration frequency must be synchronized with the frequency of the flashing relay. If one blinker light is damaged, the frequency of haptic feedback has to be doubled – as it is done by the original relay.

Tactor Placement for Skin Stimulation

There is evidence that the density of cutaneous mechanoreceptors is not uniformly distributed around the body [424]. The density values for both back and bottom allow an estimation of the optimal number and distance of actuators to be integrated into a *vibro-tactile seat.* Table 12.6 exhibits the relevant part of a two-point discrimination threshold experiment (adapted from Kandal *et al.* [12], Gallace [13] and Gibson [14], [15]).

Body Part	Threshold Distance	Tactors/Dimension
Back	$39mm$	$5.52 \rightarrow 6$
Thigh	$42mm$	$5.12 \rightarrow 6$
Belly	$30mm$	$7.17 \rightarrow 7$
Shoulder	$41mm$	$5.24 \rightarrow 6$
Upper Arm	$39mm$	$5.52 \rightarrow 6$

Table 12.6: Threshold distances and resultant number of tactor elements for several parts of the body, relevant for in-seat sensing.

Implicit Feedback

The concept of a vehicle seat as proposed here, equipped with both *vibro-tactile elements* and *pressure-sensitive sensors*, demonstrates its potential: Driver notification (stimulation) would operate fully implicitly, requiring no calibration or adjustments regarding a driver's individual characteristics, such as size, weight or even figure.

Pressure-sensing studies, as conducted in the experiments 1 to 3 for instance, showed that a person-dependent sitting posture (collected on the seat and back mats) is not static during a test run, but rather highly dynamic (caused e. g. by bumpy roads, curbs, or road holes, or originating from lateral forces), which makes it difficult to transmit directed vibro-tactile notifications to the driver because his/her position on the seat varies during the journey.

The utilization of force sensor arrays allows this issue to be compensated for by dynamically sensing both the sitting position and the orientation of the driver, and adapting the vibro-tactile feedback patterns accordingly. This approach should ensure proper feedback in any driving situation and for every driver (Figure 12.18 highlights the idea).

12.3.3 Requirements Analysis

Haptic feedback content with proper tactile cues has the potential to make overall interaction more ergonomic and natural – but on the other hand, providing users with inadequate feedback might impair their performance [130]. When intending to use haptic stimulation as an additional feedback dimension, a number of parameters have to be taken into account [425, pp. 2], e. g. (*i*) activation frequency and duration, (*ii*) intensity range and recovery times, (*iii*) kind of stimulation device, e. g. pneumatic, thermal, piezo-electric, vibro-tactile, (*iv*) complex stimulation waveforms[74], and (*v*) actuator location and distance.

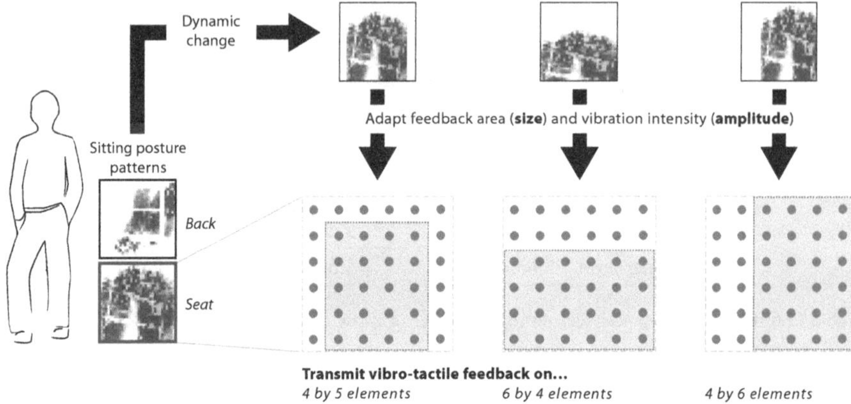

Fig. 12.18: Mapping of feedback regions according to a driver's sitting position on the vibro-tactile seat.

In this work, investigation is limited to the attributes *location* and *position* as specified by the threshold distances indicated in Table 12.6. Empirical evaluations regarding these criteria are carried out with a prototypic *vibro-tactile* seat system. Before transmitting haptic stimulations to the driver it has to be ensured that he/she is able to realize the full range of haptic information in the designated manner – this can be easily verified by evaluating pressure maps captured with force sensor arrays in the seat.

The findings from this study are mechanisms for dynamically reconfiguring vibro-tactile patterns delivered to the *vibro-tactile* seat. The proposed solution assures this by (*i*) stretching or compressing tactile patterns in both x and y axis according to a driver's actual sitting posture, and (*ii*) adapting the vibration amplitude in the z-axis in accordance with to the driver's estimated weight (separately for the seat and back subsystems).

Specifications of the Vibro-Tactile Seat

The seat utilized in the prototype has an interaction area of 430*mm* by 430*mm* and is an extension of the vehicle sensing system built in 2007. Posture patterns of the driving person are captured from force sensor arrays integrated into both the seat and backrest of the driver seat.

[74]In this work this parameter is referred to as the concept of *tactograms*.

These (input) sensors offer the potential for limited person identification [426] or, when evaluated dynamically, allow the execution of driver-influenced implicit steering activities [421].

Aside from sensing, the *vibro-tactile seat* provides haptic output capabilities. The value of haptic feedback in a real driver seat has been evaluated, e. g., in [427]. Considering the relevant threshold distances from Table 12.6, the seat and backrest should be equipped with a square matrix of 6 by 6 tactor elements each[75] (see equation (12.11) or Fig. 12.19).

$$\frac{430mm \text{ [seat size]}}{2 \times 39mm \text{ [threshold } \varnothing]} = 5.52 \text{ [tactors/dimension]} \tag{12.11}$$

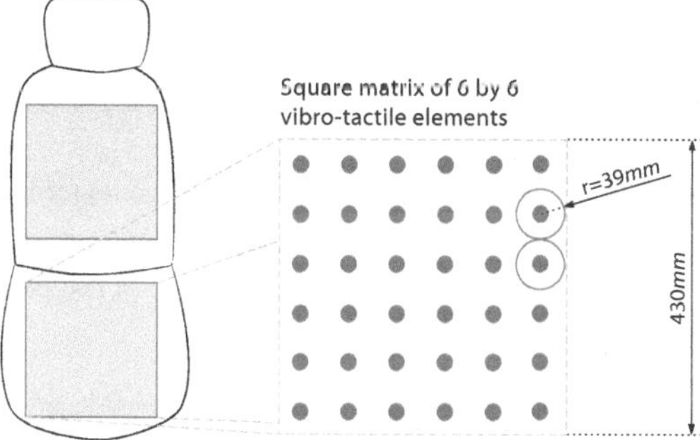

Fig. 12.19: Layout of the vibro-tactile feedback component for a vehicle seat (with respect to the relevant threshold distance).

Utilization Tuples

For characterizing the currently covered regions on the pressure mats, two 4-tuples, one for the seat and one for the back, were utilized (equations (12.12), (12.13)).

$$(L,R,B,F),(L,R,U,D) \tag{12.12}$$

$$1 \leq L \leq R \leq 32;\ 1 \leq B(U) \leq F(D) \leq 32 \tag{12.13}$$

L*(eft)* is the index of the first occupied column, R*(ight)* is the index of the last used column, B*(ack)* indicates the index of the first allocated row on the back of the seat, and F*(ront)* is the number of the last used row (on the front side) of the seat. U*(p)* and D*(own)* are the parameters on the back mat, corresponding to "B" and "F" from the seat mat. The utilization tuple for the example in Figure 12.20 would be, for instance

$$(L,R,B,F) = (3,32,2,32) \tag{12.14}$$

[75] The calculation of the required number as well as the position of tactor elements based on these threshold values is presented in [428, p. 3].

The total area of one mat is calculated as $A_T = 32$ rows $\times$ 32 columns $= 1,024$ (sensor elements). The definition of the covered area A_C is given in equation (12.15), with a maximum of A_C $(= A_T) = 1,024$ if the entire mat is covered.

$$\begin{aligned} A_C &= (R - L + 1) \times (F - B + 1) \\ &= (32 - 3 + 1) \times (32 - 2 + 1) = 930 \end{aligned} \tag{12.15}$$

$$\begin{aligned} A_{C,\%} &= \frac{A_C \text{ [Covered Area]}}{A_T \text{ [Total Area]}} \times 100\% \\ &= \frac{930}{1,024} \times 100\% = 90.82\% \end{aligned} \tag{12.16}$$

12.3.4 Experimental Results

For an empirical evaluation of the effective benefit of a vibro-tactile feedback system, dynamically modifying haptic feedback based on the current sitting posture of the driving person, several experiments were conducted on a total number of $N = 34$ test persons (voluntary students, colleagues, and friends).

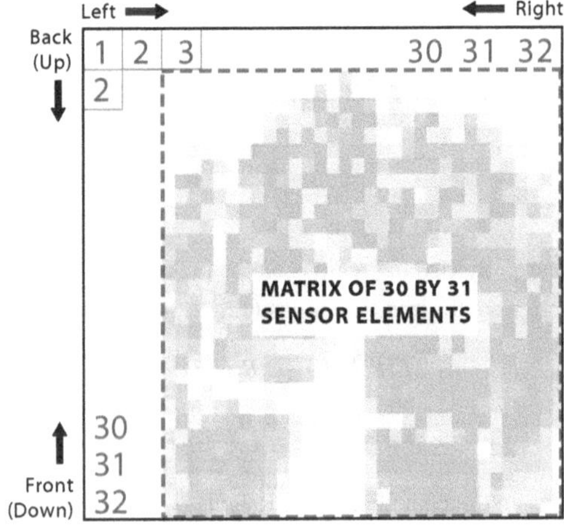

Fig. 12.20: The entire pressure sensing area on the seat mat, and the part currently covered by the driver (dashed frame).

Statistics on Test Attendees

A total of 34 test persons participated in the experiment, 27 male (= 80%) and 7 female (= 20%) subjects. Male subjects varied in size from $s_{min} = 168cm$ to $s_{max} = 194cm$ ($\overline{s} = 179.44cm$,

$\sigma_s = 8.74cm$), and in weight from $w_{min} = 58kg$ to $w_{max} = 104kg$ ($\overline{w} = 75.63kg$, $\sigma_w = 10.61kg$). Female test persons varied in size from $s_{min} = 160cm$ to $s_{max} = 173cm$ ($\overline{s} = 165.86cm$, $\sigma_s = 5.67cm$), and in weight from $w_{min} = 46kg$ to $w_{max} = 70kg$ ($\overline{w} = 58.00kg$, $\sigma_w = 5.69kg$).

Each of the test attendees was asked to adjust the vehicle seat to sit in a comfortable driving position, with both hands on the steering wheel and feet reaching clutch and brake pedals. Moreover, they were instructed to direct their point of view through the front windshield toward the road. All artefacts in their back pockets (such as a wallet or a bunch of keys) had been removed beforehand and the angle between the seat and back had been fixed for the whole experiment to guarantee data consistency.

Pressure Sensor Readings

For the first series of experiments, only static sitting postures were considered. The experiment was conducted inside a comfort station wagon with pressure sensor mats attached to both the seat and backrest. In order to accomodate sudden movements and avoid measurement errors during data acquisition, the measuring process was accomplished in two stages. First, a series of consecutive sensor readings was recorded and the median for each of the $1,024$ sensor values was calculated and stored. In a second step, the first stage was repeated at least four times[76] so that the database finally contained four complete pressure pattern sets for each test participant.

Statistics on Mat Coverage

The mean represents the number of sensors on a specific mat occupied on average by all 34 test participants and was calculated for the two threshold levels 5%[77] and 10%. The mean values of mat coverage as shown in Tables 12.7 and 12.8 are each considerably below the total mat area (although only closed, rectangle regions had been taken into account). The results of mat coverage are more expressive if indicated as percentage values.

The occupation for a threshold of 5% was between 76.17% and 96.88% for the seat mat, and between 43.07% and 93.75% for the back mat. The calculations of the corresponding mean values (90.64%, 66.50%) are given in equations (12.17) and (12.18).

$$\overline{A_{C,\%}, \text{Seat}_{5\%}} = \frac{928.147}{1,024} \times 100\% = 90.639\% \tag{12.17}$$

$$\overline{A_{C,\%}, \text{Back}_{5\%}} = \frac{680.971}{1,024} \times 100\% = 66.501\% \tag{12.18}$$

Results from both the 5% and 10% case of this evaluation confirmed the assumption that it is necessary to recalibrate tactile output in vehicles in order to guarantee universal and similar vibro-tactile information for every driver.

[76] The value "four" was coosen in accordance with the Hamming distance.

[77] A threshold of 5% means that a particular sensor is included if its pressure value is 5% of the maximum pressure value captured on the mat, or above.

Calibration Options

I All-Purpose (one size and position for all the drivers): Using this first, simplest option would result in a small-sized and inexpensive vibro-tactile feedback system, set up under consideration of the maximum regions occupied by all test persons (these are specified by the two tuples stated in equations (12.19), (12.20)). For determining the parameters of this all-purpose type, the more stable threshold level of 10% was used (see Tables 12.7, 12.8).

Statistical Value	Front	Back	Left	Right	Area
Sensor pressure (weight) > 5% of maximum mat pressure					
Physical limit	1	32	1	32	1,024
Min x_{min}	1	31	2	30	780
Max x_{max}	7	32	7	32	992
Median $\widetilde{x}$	1	32	3	32	960
Mean $\overline{x}$	1.882	31.941	2.970	31.853	928.147
Standard dev. σ	1.472	0.239	1.167	0.500	62.46
Sensor pressure (weight) > 10% of maximum mat pressure					
Physical limit	1	32	1	32	1,024
Min x_{min}	1	30	2	29	702
Max x_{max}	7	32	7	32	992
Median $\widetilde{x}$	1.500	32	3	32	914
Mean $\overline{x}$	2.265	31.765	3.265	31.529	893.176
Standard dev. σ	1.763	0.554	1.136	0.928	81.276

Table 12.7: Statistics on mat coverage for the seat mat (population N=34, two thresholds 5%, 10%). Directions according to Fig. 12.20.

These two areas, covering 552 sensors (seat) and 192 sensors (back) (also shown as the planar plateaus at height 34 in Figures 12.21 and 12.22, are significantly smaller than the regions of any individual, ranging from 780 to 992 for the seat mat and from 441 to 960 for the back mat (threshold level 10%). When activating tactor elements only within these two regions, each driver should receive all possible vibro-tactile patterns and have the ability to interpret them in the designated meaning.

$$(L,R,B,F) = (7,29,7,30);$$
$$A_{C,Seat} = (29-7+1) \times (30-7+1) = 552 \quad (12.19)$$

$$(L,R,U,D) = (9,24,11,22);$$
$$A_{C,Back} = (29-4+1)\times(22-11+1) = 192 \quad (12.20)$$

Statistical Value	Up	Down	Left	Right	Area
Sensor pressure (weight) > 5% of maximum mat pressure					
Physical limit	1	32	1	32	1,024
Min x_{min}	1	22	1	24	441
Max x_{max}	10	32	9	32	960
Median $\widetilde{x}$	1	31	6	27	644
Mean $\overline{x}$	2.457	30.286	4.971	27.829	680.971
Standard dev. σ	2.381	2.334	2.606	2.162	141.714
Sensor pressure (weight) > 10% of maximum mat pressure					
Physical limit	1	32	1	32	1,024
Min x_{min}	1	22	1	24	441
Max x_{max}	11	32	9	32	928
Median $\widetilde{x}$	1	30	6	27	601.500
Mean $\overline{x}$	2.686	29.657	5.314	27.286	635.118
Standard dev. σ	2.587	2.413	2.665	1.872	135.118

Table 12.8: Mat coverage statistics for the back mat (population N=34, two thresholds 5%, 10%).

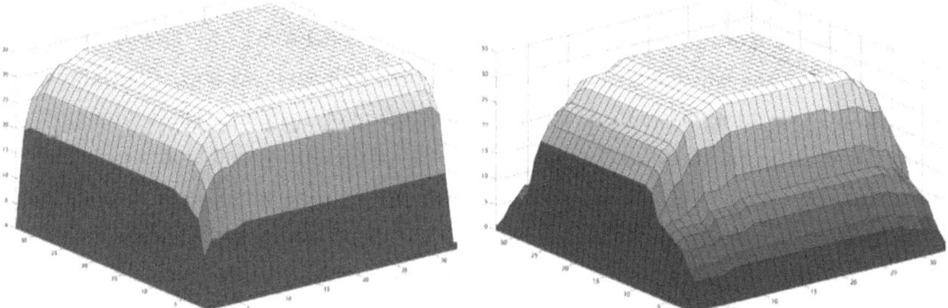

Fig. 12.21: Mat coverage on the seat. The planar plateau at height 34 is that region on the seat covered by all test candidates.

Fig. 12.22: Contact with the back mat for 34 subjects. The plateau size is significantly smaller than that for the seat (see Fig. 12.21 on the left).

A small-sized and therefore inexpensive general purpose solution for vibro-tactile stimulation could be manufactured, equipped with only 4 by 4 elements on the seat and 4 by 2 tactors on the back[78] (as indicated in equations 12.21 to 12.24), which would be a reduction compared to the initial system of 6 by 6 elements to $((4\times4)/(6\times6)+(4\times2)/(6\times6))\times100\% = 33.33\%$.

No further reconfiguration or calculation at runtime would be necessary – the small area of this system should always be covered by (driving) persons of any figure and in all driving situations.

$$\begin{aligned} T_{Seat,x} &= \frac{L-R+1}{\text{sensors}_x} \times \text{actuators}_x \\ &= \frac{29-7+1}{32} \times 5.52 = 3.967 \rightarrow 4 \end{aligned} \tag{12.21}$$

$$\begin{aligned} T_{Seat,y} &= \frac{F-B+1}{\text{sensors}_y} \times \text{actuators}_y \\ &= \frac{30-7+1}{32} \times 5.52 = 4.140 \rightarrow 4 \end{aligned} \tag{12.22}$$

$$\begin{aligned} T_{Back,x} &= \frac{L-R+1}{\text{sensors}_x} \times \text{actuators}_x \\ &= \frac{29-4+1}{32} \times 5.52 = 4.485 \rightarrow 4 \end{aligned} \tag{12.23}$$

$$\begin{aligned} T_{Back,y} &= \frac{D-U+1}{\text{sensors}_y} \times \text{actuators}_y \\ &= \frac{22-11+1}{32} \times 5.52 = 2.07 \rightarrow 2 \end{aligned} \tag{12.24}$$

II Static (individual number and position of active tactors per driver): A *vibro-tactile seat system* using this second calibration option would operate already user-specific by employing the personal static sitting postures of the driver at the time of boarding. These sitting posture images could either be added to the personal profile of the driver, and stored in the car, or recaptured every time he/she gets into the car. In the latter case, the optimal region for the haptic feedback would be calculated on-the-fly, incorporating slight changes in the sitting behavior of a person from time to time, caused, for instance, by different clothes or a changed mental state.

The resulting rectangular regions for the seat and backrest act as a *size-restricting filter* for the vibro-tactile output areas (maximum of 6 by 6 tactor elements per mat), and are fixed for the whole journey (see Figures 12.18, 12.23). These feedback regions were calculated for the sitting postures of all 34 test persons – the areas applicable for (vibro-tactile) feedback are between 76.17% and 96.87% of the entire mat size (calculation for the seat mat and the 5% threshold

[78] In order to guarantee total perceptibility by everyone, all values must be rounded off (only the value of 3.967 in equation 12.21 has been rounded up due to its close vicinity to the next integer).

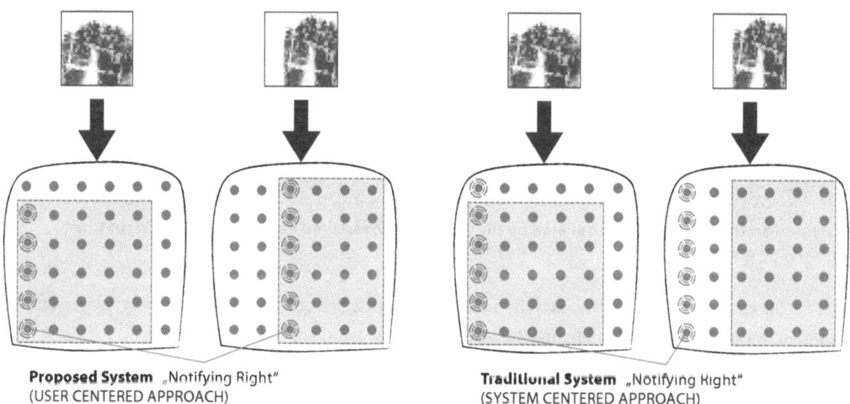

Fig. 12.23: Mapping strategies for haptic feedback according to driver's sitting behavior.

case); for more details see the subsection "Statistics on Mat Coverage" above. The evaluation and assignment of suitable feedback regions only once, at the time of boarding, is suboptimal, because of unconsidered, but substantiated, movements of the driver on longer journeys (known as "see-sawing"). However, option II would be much better than case I.

No matter how small the current feedback areas for a specific driver are, in order to guarantee universal applicability, the entire system of 6 by 6 vibro-tactile elements has to be integrated all the time.

III Dynamic (feedback area adjusted in real-time): The third option is an extension of case "II Static" and is the most complex setting. It is characterized by a dynamic, real-time calibration process, continuously adapting the vibro-tactile feedback areas according to a driver's actual sitting behavior.

Accordingly, the coverage of both pressure-sensitive mats is evaluated constantly and dynamically affects the position as well as the amplitude of vibro-tactile feedback (as outlined in Figure 12.24).

12.3.5 General Findings

This research study was focused on the development of an innovative *vibro-tactile seat* system for providing vibro-tactile feedback. Consequently, two types of mats, one for pressure sensing and the second for haptic stimulation, similar in size and shape, were integrated into the seat and backrest of a vehicle seat. The novelty in the proposed configuration is that the size and place

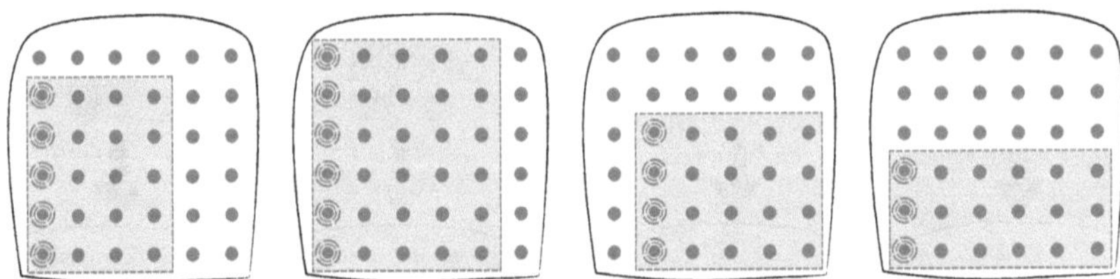

Fig. 12.24: Haptic patterns for the activity *turn right* (from a driver's view), dynamically adapted to the region of the seat currently used by the driver (lightgray rectangles).

of vibro-tactile notifications are reconfigured based on a driver's sitting posture by stretching or compressing tactile patterns in the x, y, and z axes.

Empirical evaluation of a large number of pressure data sets endorsed the expressed assumption that there is a demand for reconfiguring the regions of haptic feedback in order to guarantee similar perception of a certain vibro-tactile pattern for any driver and in any driving situation.

The described experiment relates to **Hypothesis I** (Input, e. g. driver identification or activity recognition) and **Hypothesis II** (Output, such as vibro-tactile notifications for supporting visual and auditory sensory channels) and applies the findings from the previous experiments 1 to 3 to improve and generalize vibro-tactile feedback (investigated in the following experiments 5 and 6).

Its goal was not to directly support one of the **Subhypotheses II.i** to **II.iv**, but rather to generally highlight the applicability and/or usability of vibro-tactile feedback in highly dynamic environments, such as the automotive domain.

12.4 Simulating Real-Driving Performance

Experiment 5, 6

The problems of high cognitive load and/or distraction addressed in these experiments originate from (*i*) technological advances and a rising number of Advanced Driver Assistance Systems (ADAS) in vehicles, (*ii*) physical limitations of the typically utilized sensory modalities *vision* and *hearing*, and (*iii*) personal characteristics (*age*, *gender*) of the driver, as stated repeatedly in this research work.

Considering these issues, the goal was to evaluate the applicability of the sense of *touch* as an additional information channel for Vehicle-2-Driver (V2D) feedback compared to *vision* and *hearing* with respect to the factor *response time* (or reaction performance). A constitutive experiment regarding age and gender-related constraints in vehicle operation was designed to investigate the impact of these personal attributes on the sensory channels vision, sound, and touch.

Reasons for Simulation
The evaluation of human response time in the automotive domain is normally carried out using real driving experiments. On-the-road studies are a well-established approach for investigating the usability and performance of Driver-2-Vehicle interfaces, and are furthermore renowned for providing realistic results.

These potential benefits are faced with numerous drawbacks affecting real studies, such as (*i*) high costs in terms of time and money (vehicle, detours, road block), (*ii*) dependency on environmental conditions (weather, visibility, traffic density, etc.), or (*iii*) a lot of time required for conducting the experiment. The most influential disadvantage is the high risk of danger for the test persons with on-the-road studies of user interfaces (particularly experiments with new interface designs often lead to an excessive user demand and to operation errors).

In view of these issues, a trace-driven simulation approach was chosen for this series of experiments. Apart from the required safety for test attendees (simulation would avoid situations of possibly fatal danger and would furthermore secure test participants as well as other road users from casualties or injuries), this approach generates a number of additional benefits, such as (*i*) repeatability (any experiment can be reproduced with exactly the same settings, for instance, in order to verify uncommon results), (*ii*) equality (data sets can be directly compared with each other due to exactly the same experimental flow; this would not be possible for tests in real environments, because of the high road dynamics), and (*iii*) universality (trace-driven simulation can be performed at any place and time, and independent of environmental conditions).

12.4.1 Distraction Classes

This subsection gives a short overview of the different distraction classes affecting vehicle handling, considered for the conducted experiments.

Interaction in Vehicles

The effect of cognitive (over)load with respect to the increasing number and complexity of ADAS is intensified due to the fact that the class of driving-independent devices (multimedia equipment, audio or video-player, Internet consoles, hand-held or hands-free cell phones [429], etc.) in the car is growing and additionally demands a driver's attention. For instance, there is evidence that a cell phone call requires the driver to turn his/her view from the roadside to the display while dialing, and furthermore requires his/her attention while talking to a dialog partner [430]. Both of these annoyance factors lead to a performance deterioration as just recently confirmed by E. Vingilis, a member of the the Canadian Association of Road Safety Professionals[79].

Physical Limitations

The sensory channels *vision* and *hearing* are subject to physical limitations affecting the interaction performance. Examples of restrictions for the auditory sense are environmental noise, communication between passengers, or loud music from the car stereo (for further information see the paragraph "State of the Art in In-Car Interaction" on p. 41). The visual channel is influenced by factors like visibility conditions, day and night vision capabilities, lighting conditions, or the time required by the eye for accommodation (see p. 72 or 73 for additional information). The significance of glance times in this context has been evaluated, e. g., by Green [431]. He found that glances of 2 seconds lead to 3.6 times more lane departures than glances of only 1 second would do.

Characteristics of a person, such as his/her *age* or *gender*, are an additional factor influencing reaction speed and mode.

Age Dependency: Human-Vehicle Interaction (HVI) (or more general HCI) is affected by the driver's age across all the modalities *seeing*, *hearing*, and *touch*.

Particularly with regard to the advancing age of the population, this factor is becoming of increasing importance. A literature study concering the deterioration with increasing age is given in Appendix "Age Dependency" on p. 227 and allows the following conclusion: There is

[79]Evelyn Vingilis, auto safety expert and professor at the University of Western Ontario's department of family medicine, has declared that a total prohibition of cell phone activity is needed, and has explained this with the statement "[..] if you are dividing your attention, something's going to be short-changed [the attention to the task of driving]." [101].

evidence that the accuracy of stimuli perception is affected by the individual's age. For example, for haptic stimuli it was determined by Brammer *et al.* [432] that threshold mediated by the Pacinian mechanoreceptors increases $2.6dB$ per 10 years. This has been confirmed by Shaffer and Harrison [433] and Fryer [195, p. 28]) who found that human PC decrease in number with advanced age, vibro-tactile sensitivity becomes impaired with age, and older adults required significantly greater amplitudes of vibration to achieve the same sensation-perceived magnitude as younger subjects. Likewise, Smither *et al.* [434] found that older people experience a decline in the sensitivity of skin, and also have more difficulty in discriminating shapes and textures by touch.

Analyzing the visual und auditory senses, a detected declining speed of information processing in conjunction with an increased response time leads to a drastic age-proportional increase in traffic accidents (e. g. hitting the brake pedal of a car when the traffic light turns to red) (Kallus *et al.* [435].).

When the relationship between age and the factors of deterioration is known it would be perhaps possible to compensate for them, at least partly. For instance, for the sense of touch, the vibration amplitude may be adjusted by an age-dependent factor, so that vibrations will be perceived in a similar manner by persons of any age. Furthermore, it has been determined that experience with a specific task apparently compensates for the age-influenced declining information processing speed (Breytspraak [436]).

Gender Dependency: As the female and male population is almost uniformly distributed all over the world[80], a corresponding influence on response time and/or reaction accuracy needs to be taken into account (detailed information regarding the sex ratio is given in the Appendix on p. 229).

It has been substantiated, for instance, that in almost every age group and across the different interaction channels men react more quickly than women, while women were more accurate (Adam *et al.* [437], Dane *et al.* [438], Der and Deary [439]), Speck *et al.* [440], Barral *et al.* [441], Bellis [442], Engel *et al.* [443]).

With respect to these results, it would be feasible, for example, to select the behavior of automotive user interfaces separately for female and male drivers by assigning the mode of operation automatically according to the detected driver. A more detailed investigation with respect to sex dependency is presented in the Appendix "Gender Difference" on p. 228.

[80]This is particularly true for the population of 20 to 64 year olds in the western industrialized countries, where the sex ratio is 98.8 (midyear 2008) [22].

12.4.2 Experimental Design

All the tests were carried out in a stopped car, utilizing, apart from *vision* and *hearing*, the sense of *touch*[81] as an additional information channel. Haptics has the potential to assist today's permanently charged eyes and ears, and furthermore, to improve Driver-2-Vehicle (D2V) interaction, as, for instance, confirmed by Bengtsson *et al.* [444], Amditis *et al.* [445], Harrar and Harris [446], Jones *et al.* [447], or Ho *et al.* [216].

The main goal of this series of experiments was a comparison of the adequacy and accuracy of vibro-tactile stimulations in contrast to vision and hearing. Particularly of interest were the questions if (*i*) the response time for haptic stimulation is similar to a response to heard or seen notifications and, furthermore, (*ii*) the response time decreases with the progress of the experiment as a result of, e. g. familarity with the test environment, and (*iii*) the performance (and even sequence) of the different modalities used for feedback is person dependent.

Subsequently, an overview of the automotive simulator design is presented. An in-depth description of the conducted experiments, including evaluation and results, can be found in Riener *et al.* [427] and Riener [403].

Taping

A driving scenario through the city of Linz, Austria, with (*i*) controlled and (*ii*) uncontrolled crossings, (*iii*) road tunnels and (*iv*) freeway components was recorded with a digital video camera mounted on the windshield of a car. The waiting times at crossings and unsubstantial or pointless driving sections were later deleted from the taped run so that the video ultimately had a length of 11*min*. 22*sec*.

Tagging and Instruction Editor

The final cut video was integrated into the simulation application using the Java Media Framework (JMF) in version 2.1.1e. In the timeline, 44 points were tagged as positions for the later triggering of specific user actions. Only *valid actions* were tagged (e. g. a test participant could not receive a "turn left"-request in a situation where left turns are impossible in the corresponding line of the road). In the current prototype only the four activities (*i*) turn left, (*ii*) turn right, (*iii*) low-beam lights on, and (*iv*) low-beam lights off were differentiated and evaluated – to each of these activities a visual, auditory, and haptic signal was assigned (as described in Table 12.9).

To simplify the assignment task among the 44 vehicle control instructions and the synchronized points in the video, the software component "Instruction Editor" was implemented. It

[81] The majority of information is delivered via visual and auditory senses, followed by the tactile channel with a contribution level of up to 10% (an introduction into information bandwidth and other characteristics of individual sensory channels is given in the section "Perception and Articulation" on p. 27).

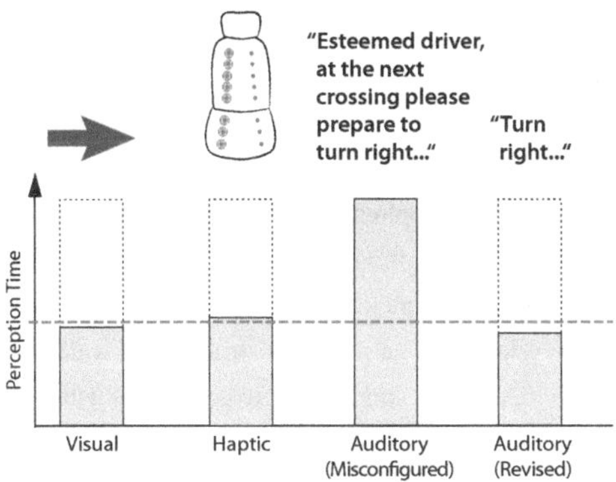

Fig. 12.25: Perception times of individual feedback channels have to be aligned to each other in order to get meaningful results.

Signal	Visual	Auditory	Haptic
Turn Left	Symbol "Left" (Superimposed on the video)	"Turn Left..." (Spoken instruction)	All 8 left are activated simultaneously
Turn Right	Symbol "Right" (Superimposed on the vidco)	"Turn Right..." (Spoken instruction)	All 8 right are activated simultaneously
Lights On*)	Symbol "Lights on" (Superimposed on the video)	"Lights On..." (Spoken instruction)	All 6 tactors on the seat are oscillating
Lights Off*)	Symbol "Lights off" (Superimposed on the video)	"Lights Off..." (Spoken instruction)	All 6 tactors on the seat are oscillating
*) *The binary item "light switch" uses the same haptic patterns for on and off.*			

Table 12.9: Evaluated activities and corresponding feedback signals.

allows the definition, modification, and storage of individual instruction sets per video in an *.eli*-file (=**e**vent **li**st). For the current *trace driven simulation* experiment this editor was only used to inspect and adjust the individual trace lines (which were recorded during the taping journey).

The parameter list of the instruction editor is extendable; up to now visual, auditory, and haptic notifications can be selected and assigned. The duration of user notifications can either be set individually or assigned automatically by the software. Moreover, it is possible to choose a specific interaction modality for each instruction point (e. g., to evaluate a whole trace only with haptic instructions) or let the application itself select one randomly (preferred option). Corresponding to the behavior of the modality, different visuals, audio files, or vibro-tactile patterns, together with additional parameters such as vibration amplitude, frequency, duration, etc. can be defined and/or selected. The principal structure as well as possible parameters of an *.eli*-file (event list), together with a short example composed using the instruction editor, is shown in Fig. 12.26.

```
# Assigned IDs
0...Visual Task
1...Auditory Task
2...Haptic Task
3...Random Task (one out of 0, 1 or 2)

# Structure of individual tasks (0,1,2,3)
ID;Task Name;Stop Key;Trigger Time (ms);Serial ID;Image File;Label
ID;Task Name;Stop Key;Trigger Time (ms);Serial ID;Sound File
ID;Task Name;Stop Key;Trigger Time (ms);Serial ID;Touch File
ID;Task Name;Stop Key;Trigger Time (ms);Serial ID;Image File;Sound File;Touch File;Label

# Examples (single tasks only)
0;Right1;s;9484;Turn Right;C:\\driveSim\\images\\turnRight.jpg;right
2;Left1;a;22030;Turn Left;C:\\driveSim\\haptics\\turnLeft.bis
1;Right2;s;69722;Turn Right;C:\\driveSim\\sound\\turnRight.wav
```

Fig. 12.26: Valid task identifiers and their parameters.

Sequence Creator

The sequence creator is a toolset for defining and organizing individual vibro-tactile patterns (so-called "tactograms"). A "*tactogram*" specifies the dynamic behavior of all vibro-tactile elements (*tactors*) in a system. *tactograms* can be loaded from the file system, changed, and stored again. Pattern files are indicated by the file extension *.bis* (=**b**oard **i**nstruction **s**et). This application is predestined to be used as a rapid prototyping tool for haptic patterns. Individual templates could be defined, allowing variations by adapting the following parameters: (*i*) Arbitrary selection of tactors at each step in the instruction set, (*ii*) selection of vibration frequency in the full tactor range ($10Hz$-steps from 10 to $2,500Hz$, nominal center of $250Hz$), (*iii*) selection of discrete gain levels (0, 1, 2, and 3), and (*iv*) freely configurable activation and pause periods in ms-resolution. There is no limit with respect to the length and complexity of an instruction list.

A set of instructions defined (or loaded) on-the-fly can be directly transmitted to the tactor system and evaluated instantly. If the perception of the tested pattern is unpleasant it could be changed immediately at runtime. A feature to verify *tactograms* without a connected tactor system, simply by inspecting patterns visually on the "*visual tactor board*", has been installed as well.

Mapping

The mapping between the four activities and the corresponding visual, auditory, or haptic feedback was required to be as intuitive as possible. Another issue was the rule that each of the three signals for a specific activity had to be recognizable in approximately the same amount of time (as explained in Fig. 12.25); the final specification of the mapping is summarized in Table 12.9.

It is self-explanatory that the time for unique pattern identification would increase with a larger number of possible samples or alternatives (see for instance Testa and Dearie, [355]). In the present case the number of samples was kept constant for the entire experiment as well as across all the different feedback channels.

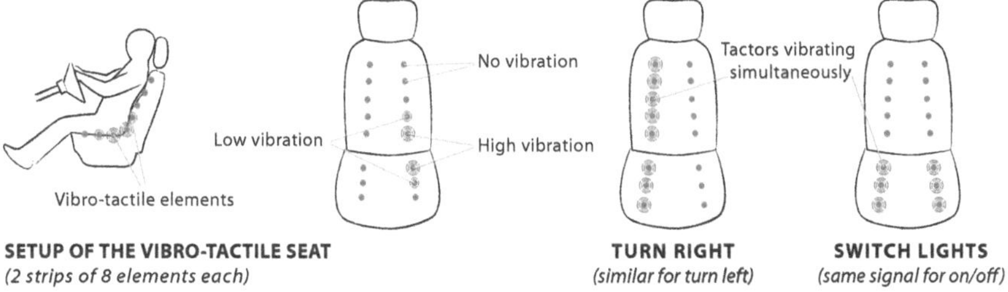

Fig. 12.27: Setup of the vibro-tactile seat used in the experiments and visual representation of two *tactograms* for right turns and switching the lights on/off.

Hardware Portion and Experiment Processing

The studies described here were conducted in an Audi 80 model parked in a single-car garage without windows to ensure high contrast for the playback of the prerecorded route (see Figures 12.28, 12.30, 12.31) and were the initial in a larger series of simulation experiments in real vehicles. The experimental system itself was designed universally so that further simulation or on-the-road tests can be conducted with the same setup (a notebook computer was used for data acquisition and the connection to sensors and actuators).

A video projector (with bright projector lamp) was mounted on the roof of the car, projecting the pre-recorded road journey on a 2 by 3 meter sized projection screen placed ahead of the front windshield, so that all the test participants could see the entire video while sitting on the driver seat (see Fig. 12.30). Auditory feedback was delivered with stereo headphones in order to prevent distractions from unintentional environmental noise. Visual instructions were

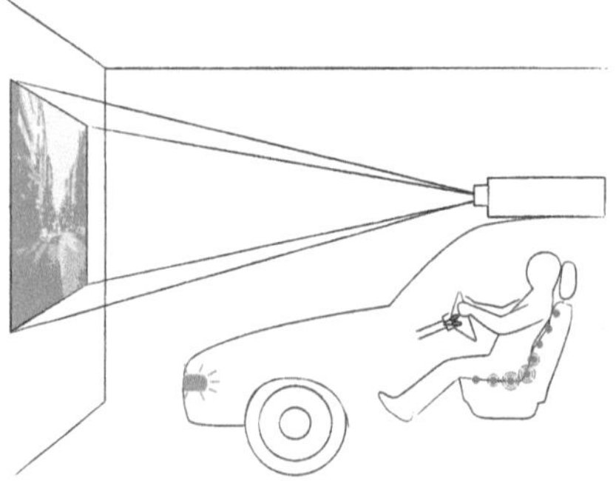

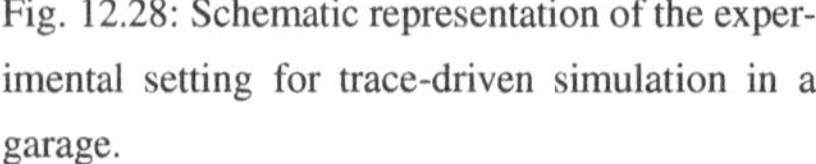

Fig. 12.28: Schematic representation of the experimental setting for trace-driven simulation in a garage.

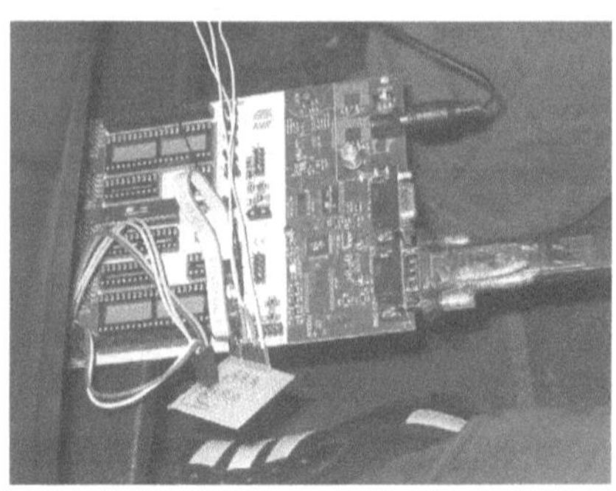

Fig. 12.29: The ATmega8-16PU microcontroller, placed on a STK500 development system, with external connectors and a voltage regulation circuit.

displayed on the projection screen, superimposed on the video of the road journey. For vibro-tactile feedback, 16 C-2 linear tactors, arranged in two strips of eight elements each, were used (see Fig. 12.27). The utilized vibro-tactile elements (*tactors*) were selected according to the required capability of providing a strong, pointlike sensation that can be easily felt and localized on the body, even through clothing.

Fig. 12.30: The garage with projection screen, data-acquisition vehicle, and processing equipment.

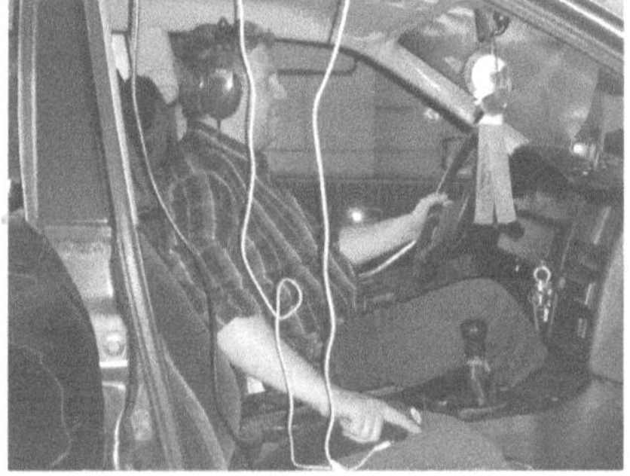

Fig. 12.31: Experiment processing inside the car (note that the test participant is equipped with headphones).

The response times (turn left, turn right, lights on, lights off) of each test participant were captured as electrical signals from the real control instruments of a car (turn indicators and headlight switch) by using an Atmel AVR ATmega8 microcontroller (the μC was placed on a

STK500 development board, and extended with a voltage regulation circuit; see Fig. 12.29), and passed over to the computer running the capturing and evaluation application. During playback of the video, the synchronized trace engine processed events, passed them over to a random generator (which selected one out of the three feedback modalities) and transmitted associated notification signals to the test person sitting in the car (using either visual, auditory, or haptic feedback). Simultaneously a timer was started, measuring the delay between one notification and the corresponding response of the test person.

The experiment procedure itself, which took about 15 minutes, was fully automated (a supervisor only ensured that the experiment was conducted correctly). For each event enqueued by the trace engine a data set containing notification time, the channel used for the feedback, the user's response time, and the switch first activated by the user (to determine if the driver had activated an incorrect switch, e. g. the "light on" switch instead of the correct "left-turn indicator" switch) was stored in a database.

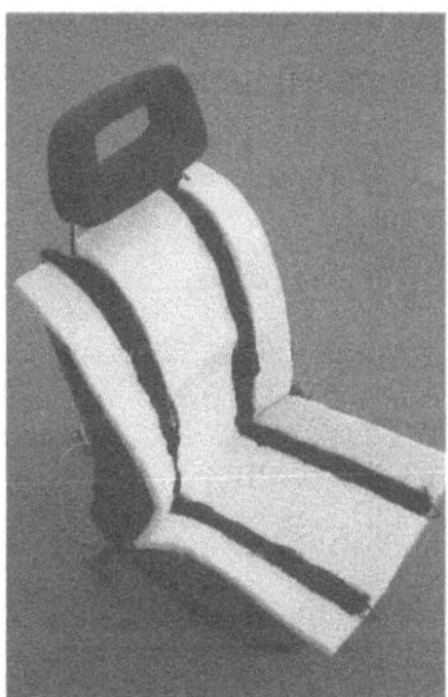

Fig. 12.32: Video frame with superimposed control window as shown in the projection (left), prototype of the vibro-tactile car seat (right).

12.4.3 Evaluation and Results

Predeterminations

Vibro-Tactile Feedback: The position of the vibro-tactile actuators in the seat as well as the parameters activation frequency and intensity (amplitude) were adjusted and fixed to provide optimal stimulation for the Pacinian corpuscles (PC).

Selection of Test Candidates: In this experiment series, the main objective was the evaluation of age- and gender-dependency on the reaction performance using different notification channels. Therefore, test participants were selected according to their age and sex. A narrow range of age and thus, a low standard deviation σ, were presumed to improve experimental results by eliminating the substantiated age-related distortions. Selecting the test participants exclusively from either the group of males or females would additionally increase the quality of acquired test data because gender-dependent differences are eliminated (e. g. response time)."

Test Participants: The experiments were conducted with 18 voluntary test persons (15 male, 3 female participants), all of them university students, research staff or friends ranging in age from 18 to 38 years (males $\varnothing 24.80 \pm 5.41$, females $\varnothing 26.00 \pm 4.00$ years) and all with a valid driver's licence. Apart from weight and size, driving experience in years was inquired during the experiment. The value was between 1 and 20 years for male participants ($\varnothing 6.93 \pm 5.19$ years) and between 4 and 12 years for female attendees ($\varnothing 8.00 \pm 4.00$ years). As each attending person got his/her driving license at the (earliest) age of 18 years, the participants' ages correlated fully with the requested driving experience. Table 12.10 gives a detailed overview of test participants' personal statistics.

Trait	Min x_{min}	Max x_{max}	Mean $\overline{x}$	Median $\widetilde{x}$	Std.Dev. σ
Male (15 subjects)					
Age	18	38	24.80	25	5.41
Weight	50	120	83.80	75	19.77
Size	170	197	180.40	179	6.25
Driving Exp.*)	1	20	6.93	5	5.19
Female (3 subjects)					
Age	22	30	26.00	26	4.00
Weight	57	75	67.33	70	9.29
Size	167	178	171.67	170	5.69
Driving Exp.*)	4	12	8.00	8	4.00
All (18 subjects)					
Age	18	38	25.00	25	5.12
Weight	50	120	81.06	75	19.29
Size	167	197	178.94	178	6.87
Driving Exp.*)	1	20	7.11	6	4.92

*) *The driving experience is given in years (whole number).*

Table 12.10: Personal statistics of experiment participants, separated for males and females.

Evaluations

For the general evaluation of response times, all the data sets originating from male test participants only were used, in order to avoid the mentioned gender-specific influence on response time.

The second examination accomplished within this series of experiments was focused on gender-related issues. The main goal of analysis was to confirm (or refute) the discerned impact of gender on response times in Human-Computer Interaction (HCI). It should be noted that the results need to be interpreted with care due to the very few available data sets from female test attendees (16.7% women compared to 83.3% men), but this was as desired and defined prior to the selection of candidates.

For the investigation into age dependency, two different approaches were evaluated, using data sets from male persons only in order to avoid a cross correlation between age and gender. First, the prerecorded data sets were separated by the median age, one group containing the younger test persons (below 25) and the other including the older participants (25 years or above). In a second analysis, age dependency was studied on data set level with the database re-sorted in ascending order by the age of test persons.

Briefing: Before starting a trace-driven journey, each test person was briefed about the requirements, expectations and goals of the experiments. After that, tests were started immediately and without a "run-in test". This possibly caused longer reaction times for the first few stimulation events and probably had an influence on the gradient of the (linear) trend lines[82] as depicted in Fig. 12.33. But on the other hand, a preceding run-in experiment would have added an unintentional learning-effect to the final recorded data and was therefore omitted.

Results

The sensory channel used for a particular notification was selected randomly and therefore indicates that the number of tasks for the three modalities was not distributed uniformly (depending on the quality of the random number generator).

I General Results: For the evaluation of reaction performance, all the data sets originating only from male test participants were used, in order to avoid the indicated gender-specific influence on response times. A later comparison of results using the entire data base, including data sets from females, revealed virtually no difference.

Fig. 12.33 shows the response times separately for each of the three notification channels utilized, vision, hearing, and touch (5% confidence interval, 628 data sets). Individual tasks are represented in the x-axis, arranged in ascending order in the timeline for all test participants.

[82]Trend lines allow for the expectation of decreasing response times as the experiment progresses.

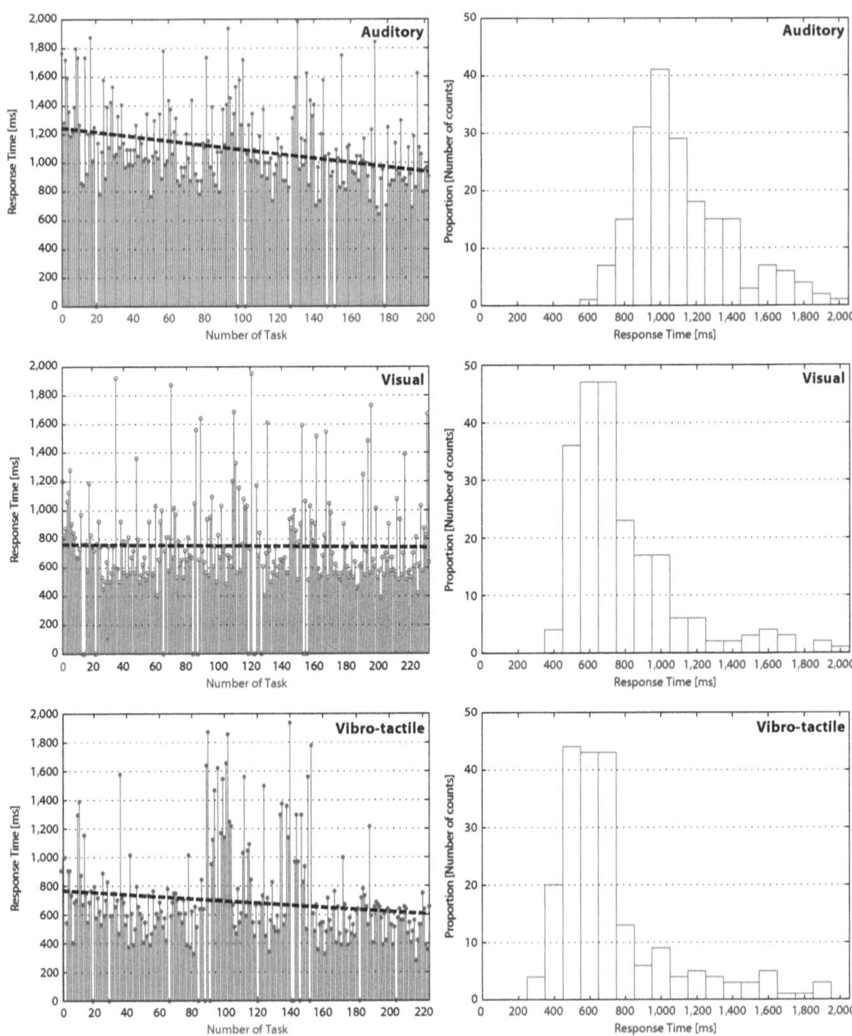

Fig. 12.33: Response times for auditory, visual and vibro-tactile stimuli (left column, from top) of male test persons only and a 5% confidence interval. The linear trend line on the response times runs downwards for all three notification modalities. The corresponding histograms in the right column show response times for the three notification modalities hearing, vision, and touch. Response to haptic notifications performs best, followed by visual and auditory sensations.

Comparing the three stem diagrams (Fig. 12.33) visually makes it clear that the fastest response was given after haptic stimulations, followed by visual and auditory notifications. Furthermore, the linear trend line of response times runs downwards for all the three modalities individually (the performance increase with regard to response time can be attributed to the familarity with the simulation environment after a while and/or the fact that learning generally leads to improved efficiency). The gradient of each of the trend lines is an indicator for the improvement achievable through training. The auditory modality offers most potential, followed by the haptic sense, with the visual sensory channel far behind, showing almost no improvement potential at all.

Trait	Min x_{min}	Max x_{max}	Mean $\overline{x}$	Median $\tilde{x}$	Std.Dev. σ
660 data sets, 3% confidence interval (=641 data sets, 97.12%)					
ALL	281*ms*	4,829*ms*	917*ms*	812*ms*	477*ms*
Visual	391*ms*	4,829*ms*	831*ms*	687*ms*	496*ms*
Auditory	641*ms*	3,532*ms*	1,134*ms*	1,055*ms*	334*ms*
Haptic	281*ms*	4,156*ms*	688*ms*	625*ms*	355*ms*
660 data sets, 5% confidence interval (=628 data sets, 95.15%)					
ALL	281*ms*	1,985*ms*	873*ms*	782*ms*	349*ms*
Visual	391*ms*	1,922*ms*	766*ms*	672*ms*	286*ms*
Auditory	641*ms*	1,938*ms*	1,111*ms*	1,047*ms*	267*ms*
Haptic	281*ms*	1,625*ms*	670*ms*	625*ms*	253*ms*

Table 12.11: General statistics on response times for two confidence intervals and all male test persons, separated by modality. User response to haptic stimuli was best, far ahead of visual and auditory stimulations.

The results on response times from the individual feedback channels as presented in Table 12.11 show significant differences in performance, for instance in the lower part of the table exhibiting the 5% confidence interval. Considering the mean reaction time, response to vibro-tactile notifications ($\overline{x_h} = 670ms$) is given 14.3% faster than to visual stimuli ($\overline{x_v} = 766ms$), and 65.8% faster in response to auditory stimulation ($\overline{x_a} = 1,111ms$).

The performance of haptics compared to feedback times for vision and hearing was very promising and encourages the application of the sense of touch for improving Human-Computer Interaction (HCI), particularly for time-critical tasks or with the aim to reduce cognitive load on auditory and visual sensory modalities.

II Gender-Related Statistics: The gender-related analysis of 18 prerecorded data sets provided the following results (it should be noted that the two groups were not uniformly distributed). The average response by male test persons was given faster than that of females; based on the mean values of all the modalities, males outperformed by $+11.3\%$ ($972ms/873ms \times 100\%$). A similar result was found for the three modalities individually, with males achieving $+10.3\%$ for auditory stimulations, $+14.1\%$ for visual stimuli and $+17.7\%$ for haptic notifications (see Table 12.12 and Fig. 12.34).

Trait	Min x_{min}	Max x_{max}	Mean $\overline{x}$	Median $\widetilde{x}$	Std.Dev. σ
Male participants, 5% confidence interval (=628 data sets, 79.29%)					
ALL	281*ms*	1,985*ms*	873*ms*	782*ms*	349*ms*
Visual	391*ms*	1,922*ms*	766*ms*	672*ms*	286*ms*
Auditory	641*ms*	1,938*ms*	1,111*ms*	1,047*ms*	267*ms*
Haptic	281*ms*	1,625*ms*	670*ms*	625*ms*	253*ms*
Female participants, 5% confidence interval (=124 data sets, 15.65%)					
ALL	500*ms*	1,984*ms*	972*ms*	883*ms*	346*ms*
Visual	547*ms*	1,828*ms*	874*ms*	734*ms*	328*ms*
Auditory	828*ms*	1,984*ms*	1,226*ms*	1,172*ms*	264*ms*
Haptic	500*ms*	1,594*ms*	789*ms*	711*ms*	252*ms*

Table 12.12: Gender-related statistics on response times, individually for the sensory channels vision, hearing, and touch (basic population of 792 data sets). Male test persons responded faster to stimuli than female participants (true for all three modalities).

The sequence of sensory modalities according to the reaction time was equal for both male and female test persons. Each of the two groups responded fastest to haptic stimulations ($\overline{x_{male,h}} = 670ms$, $\overline{x_{female,h}} = 789ms$), followed by notifications via the visual sensory channel ($\overline{x_{male,v}} = 766ms$, $\overline{x_{female,h}} = 874ms$) and the sense of hearing in last place ($\overline{x_{male,a}} = 1,111ms$, $\overline{x_{female,a}} = 1,226ms$).

III Age-Related Statistics: As there is evidence that the response time decreases with increasing age (across all the modalities and for both male and female persons), as discussed for instance in Appendix D.2 "Age Dependency" on p. 227, the selection of test participants for these series of experiments was determined by their age with a view to compensating for this issue in general response time studies. The variance among all test attendees was rather low at 25 ± 5.12 years (see Table 12.10). Additionally it has been shown, e. g. by Breytspraak

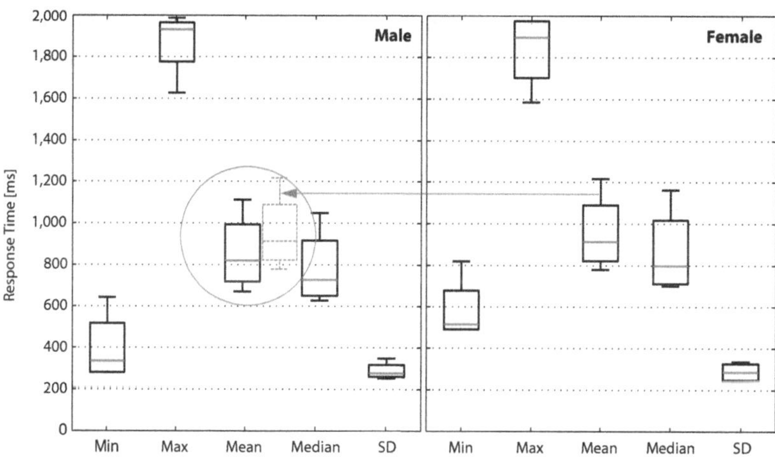

Fig. 12.34: Boxplots show differences between male (left) and female (right) test participants. A salient contrast in minimum response time in favor of males can be indicated.

Trait	Min x_{min}	Max x_{max}	Mean $\overline{x}$	Median $\widetilde{x}$	Std.Dev. σ
Younger participants, 5% confidence interval (=291 data sets, 46.34%)					
ALL	281*ms*	1,985*ms*	871*ms*	781*ms*	371*ms*
Visual	391*ms*	1,875*ms*	763*ms*	672*ms*	282*ms*
Auditory	688*ms*	1,938*ms*	1,166*ms*	1,094*ms*	288*ms*
Haptic	281*ms*	1,562*ms*	647*ms*	602*ms*	255*ms*
Older participants, 5% confidence interval (=337 data sets, 53.66%)					
ALL	328*ms*	1,953*ms*	875*ms*	797*ms*	329*ms*
Visual	406*ms*	1,922*ms*	769*ms*	687*ms*	292*ms*
Auditory	641*ms*	1,797*ms*	1,068*ms*	1,031*ms*	243*ms*
Haptic	328*ms*	1,625*ms*	691*ms*	641*ms*	250*ms*

Table 12.13: Age-related statistics for male test persons only and separated by modality show almost no difference in the response times (basic population of 792 data sets).

[436], that experience in a task (should be true at least for the visual and auditory stimulation of frequent car drivers) can eliminate the age-dependent decrease in response time (see p. 162 or p. 228).

For these reasons, the investigation of the influence of age on the response time indicated differences of only minor significance – nevertheless, prerecorded data were evaluated in order to identify possible irregularities using two approaches.

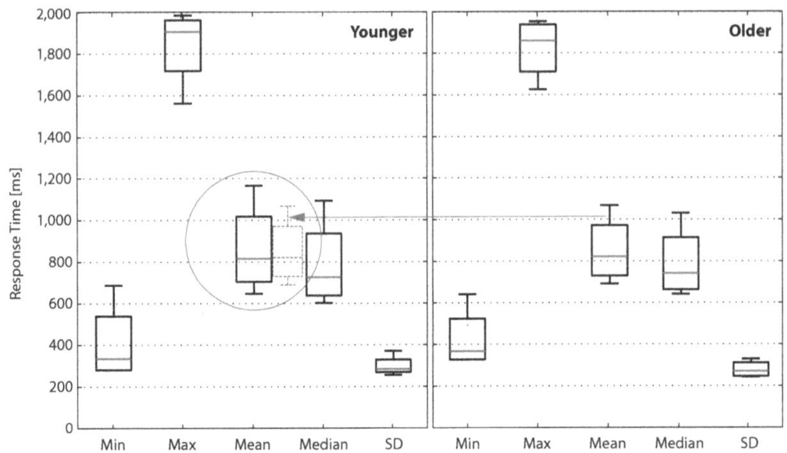

Fig. 12.35: Boxplots for younger (left) compared to older (right) test participants.

Grouping by Median: A first examination of data dependency based on age was done by grouping the data sets into two groups by the median age (below 25 years, 25 years and above). The evaluation showed that the mean response time over all modalities of the older group of males ($\overline{x_{old}} = 875ms$) compared to that of the younger group ($\overline{x_{young}} = 871ms$) was almost equal (difference of 0.5% at the expense of the older group). Considerably larger differences can be recognized when considering the three modalities individually. Deterioration with age increases to 0.9% ($\overline{x_{old,v}} = 769ms$, $\overline{x_{young,v}} = 763ms$) for visual stimulations and to 6.8% ($\overline{x_{old,h}} = 691ms$, $\overline{x_{young,h}} = 647ms$) for notifications delivered via the haptic channel. The auditory modality performs, differently as expected, most likely due to measurement errors or the low number of test persons; the older group of test participants achieved an increase in performance of 9.2% ($\overline{x_{old,a}} = 1,068ms$, $\overline{x_{young,a}} = 1,166ms$) compared to the younger group (see Fig. 12.35).

The results of this examination do not permit a statement to be made about performance deterioration with age. This was almost expected, because all of the male test persons were young with a low standard deviation *SD* of 5.12 years, and age-influenced performance deterioration is particularly anticipated and proven for older people (e. g. retired persons) or between groups of persons with large differences in their mean age (Shaffer and Harrison, [433]).

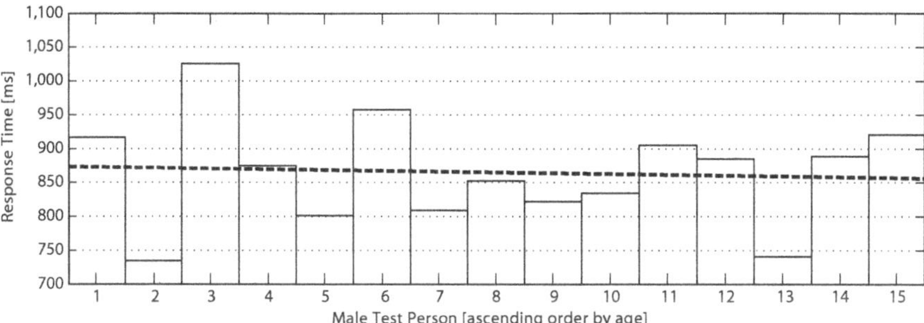

Fig. 12.36: The chart of mean response times for all modalities with superimposed linear trend line shows a slight decrease in response time (from 875*ms* to 855*ms*).

Arranged by Age: A second evaluation was conducted, considering the person's individual age instead of grouping all the test participants into two regions. For that reason, the database was re-sorted in ascending order based on the age of the (male) test persons. The chart of mean response times and the superimposed linear trend line (see Fig. 12.36) shows that the reaction performance (over all modalities) increases with rising age (the time of response decreases from approximately 875*ms* for the 18-years-old test persons to 855*ms* for the 38-years-old test participants).

The results of this experiment (the three stimulation modalities are shown separatley in Fig. 12.37) are similar to the findings of the evaluation using "grouping by median". The response time decreases with rising age for auditory (from $\approx 1,180ms$ to $\approx 1,125ms$) and haptic (from 700*ms* down to 600*ms*) stimulations, but increases for notifications using the visual sensory channel (from about 760*ms* to 900*ms*).

However, the results of the influence of a person's age on his/her driving performance are not overly significant by reason of (i) the small number of test participants (15 males), (ii) the narrow range of age, and (iii) the application of a driving simulator instead of on-the-road studies.

IV Influence of Age and Gender on Reaction Times: Considering the margins of deviation for all the experiments conducted allows the influence of age and/or gender on the reaction performance of a specific notification channel to be investigated. With respect to the 5% confidence interval, σ is at its lowest for the haptic modality in (i) the general case of male participants only (haptic $\varnothing 625 \pm 253ms$, auditory $\varnothing 1,047 \pm 267ms$, visual $\varnothing 672 \pm 286ms$), (ii) the gender-specific case for both males (haptic 625.00 ± 252.50, auditory $1,047.00 \pm 267.36$, visual 672.00 ± 286.35) and females (haptic $\varnothing 711 \pm 252ms$, auditory $\varnothing 1,172 \pm 264ms$, vi-

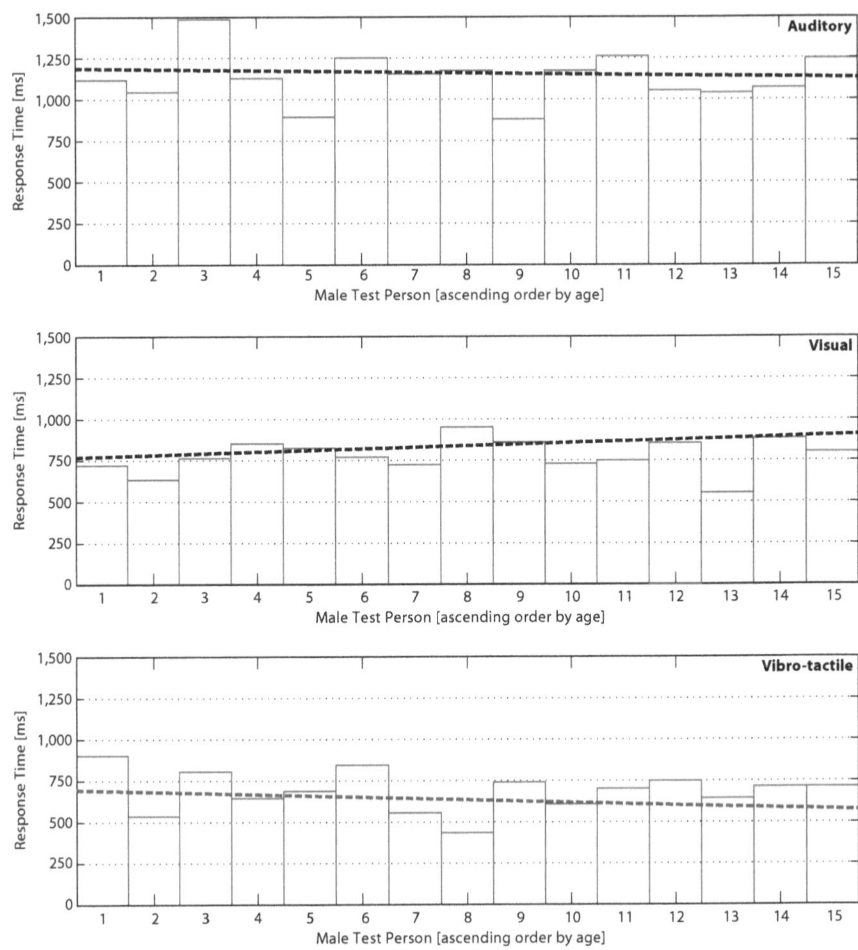

Fig. 12.37: Correlation between response time and age, separately for the individual sensory modalities.

sual ⌀$734 \pm 328ms$), and (iii) the case of the younger persons in the age-related study (haptic ⌀$602 \pm 255ms$, auditory ⌀$1,094 \pm 288ms$, visual ⌀$672 \pm 282ms$). Only for the older test participants in the age-related study did the auditory sensory modality have less deviation than the sense of touch (auditory ⌀$1,031 \pm 243ms$, haptic ⌀$642 \pm 250ms$, visual ⌀$687 \pm 292ms$), probably due to measurement errors or the low number of test persons.

12.4.4 Conclusions

In these studies, the response times (speed of reaction) when using the three different interaction modalities *vision*, *sound*, and *touch* were investigated in the automotive domain by using a trace-driven simulation approach instead of the common on-the-road studies. Apart from a general examination of the applicability of haptics as an additional feedback channel for Vehicle-2-Driver (V2D) communication compared to the established visual or auditory notification channels, the impact of age and gender on the performance of the different modalities was analyzed. In order to demonstrate a general validity of simulation results for real driving situations, these studies should be repeated with and compared to a suitable on-the-road experiment. The results of these experiments can be summarized as follows:

I Sensory Modalities

(i) **Purpose of haptic stimulation:** Simulation results confirmed the presumption that the *sense of touch* is capable of supporting in-car interaction based on the modalities *vision* or *hearing*. Measured response times when completing navigational tasks such as operating the turn indicator or light switch after a corresponding vibro-tactile impulse were rather similar to those for visually initiated or spoken ones, and often even better. In the end, vibro-tactile interfaces in vehicles can be suggested as add-ons to the established modes of interaction with the aim of reducing cognitive load or relieving the driver of distraction (which contributes to crashes in approximately 25% of the cases, see [448, p. 28]).

The simulation parameters were specified close to reality, as confirmed, for example, by the factor response time. The maximum measured value (over all modalities) was $1,985ms$ for the 5% confidence interval (see Fig. 12.11). This is just below the value of two seconds, recommended as upper limit for response in convenient Human-Computer Systems (Testa *et al.* [355, p. 63], Miller [356], or Shneiderman and Plaisant [27, Chap.11]).

(ii) **Gender dependency:** The gender-related comparison of test participants was not too significant due to the minor number of female test persons compared to the males (3 females compared to 15 males, or a rate of 16.67%).

The evaluations revealed that male test participants respond faster to stimulations than female attendees. The highest percental difference was determined with the responses to haptic notifications and was +17.66% (calculated from $788.65ms/670.30ms \times 100\%$). The sequence of sensory modalities based on the mean response times, which indicated the *haptic* channel ahead of the *visual* and *auditory* ones, was equal for both groups of test participants (female, male), and followed the order obtained in the general case considering only male participants.

(iii) **Age dependency:** A deterioration of reaction performance (or an increase in response time) with increasing age has been proved repeatedly. This fact, together with an increased life expectancy, necessitates ways and means to compensate for the impact, and furthermore establish an adaptive interface working similarly for younger and older drivers (e. g. by automatic recalibration).

The first study (male persons only, two groups separated by median age) showed almost no difference between the mean response times of the two groups (0.49%). Larger differences were indicated when comparing the sensory modalities individually. For the visual channel, the reaction performance decreased by 0.85% for the older group, compared to the younger; for the haptic channel, the determined deterioration was 6.83%. The auditory modality showed an increase in reaction performance (or a decrease in response time) of 9.18%. One potential reason for this slight disparity could be the fact that the range between minimum and maximum age (spread) was rather small. More significant results should therefore be achieved when using one group of young persons (e. g. newly licenced drivers) and another one with older people (retired persons), as for instance indicated by Shaffer *et al.* [433].

When considering the mean values of the response time of an individual over all modalities, and putting them in ascending order with respect to the age of the driver, it can be determined (for instance, by inspecting the linear trend line) that the time required for feedback decreases with increasing age, as opposed to the substantiated increase in response time with rising age. When checking the response times for each modality individually, the results are similar to those received from the first study on age dependency (decreasing response time for auditory and haptic notification channel, increase of response time for the visual modality). These results are contrary to those reported in literature, probably due to the small bandwidth of age and the fact that the evaluation was performed in a simulation environment and not within a real driving scenario. Another issue could be the already mentioned limited number of data sets available for the evaluation and the focus on male participants only.

II Trace-Driven Simulation

Results from this experiment confirm that the utilization of trace-driven simulation is viable in Human-Computer Interaction (HCI) studies. Up to now, user experiments have often been designed spontaneously, processing models with random scripts, and therefore offer no possibility of repeating them. The results from the trace-driven simulation experiments support the assumption that the class of trace-driven simulation applications has great potential in simulating HCI problems.

In order to substantiate a possible interrelationship between simulated and real studies, an on-the-road experiment considering age and gender deterioration as well as the response times for individual notifications via the *visual*, *auditory*, or *haptic* sensory modalities, has to be designed, studies have to be conducted, and the results have to be compared to those obtained in the experiments described here.

12.4.5 General Findings

Both experiment 5 (response times of sensory modalities) and experiment 6 (age and gender-related performance deterioration) provided a detailed view of the Driver-Vehicle Interaction loop with respect to the channels vision, hearing, and touch. The two experiments took drivers of different agees and sexes in clearly defined automotive applications into account.

In all evaluated test cases, the sense of touch performed (much) better than the visual or auditory channels, which supports **Subhypothesis II.i** concerning the recommendation of the delivery of important messages via the sense of touch.

Due to the experimental setting, a comparison regarding performance improvement between a system utilizing only vision and hearing and another system employing all of the three modalities vision, hearing and touch, as stated in **Subhypothesis II.i**, cannot be made.

Considering the 5% confidence interval of the various experiments allows evidence to be provided for **Subhypothesis II.ii**, as not only the reaction time, but also the margin of deviation (σ) for the haptic modality was lowest in almost every experimental setting. Only evaluation results from the older group of test participants in the age-related study ran contrary, as the sense of touch had a higher deviation than the auditory sensory modality (probably caused by measurement errors or an applied learning effect).

Elaborate evaluation of age and gender dependency showed that all of the three sensory modalities are highly dependent on both age and gender. Furthermore, the achieved results confirmed, as stated in **Subhypothesis II.iii**, that the order of sensory modalities in terms of the response time, which is the sense of touch followed by visual and auditory modalities, is affected neither by age nor by gender. All the conducted experiments showed that the utilization of the sense of touch, side by side with vision and hearing, could enhance any Driver-Vehicle Interaction system (**Subhypothesis II.iv**) and supported the integration of multimodal interfaces in vehicles, comprising, aside from the visual and the auditory notification channels, the haptic modality as a third feedback channel.

12.5 Further Experiments

Further experiments considering a more general investigation into haptic interaction and/or performance have been conducted. Their findings have been partly used in this research work (e. g. the findings of the application of different "*tactograms*" for stimulation). Nevertheless, the corresponding experiments are not presented here.

For detailed information on these studies, see, for instance, Riener *et al.* [449] or Ferscha *et al.* [211].

"Patience is bitter, but its fruit is sweet."
Aristotle (384 BC – 322 BC)
Greek critic, philosopher, physicist, and zoologist

Part V

Discussion and Conclusion

13 Predeterminations for Investigation

13.1 Domain

This research work has investigated user interfaces in the automotive domain, with a focus on implicit Driver-Vehicle Interaction. This self-contained field of research allowed experiments to be conducted with a limited number of dependent variables and constraints, and furthermore created the ability to simulate experiments with "near real-time behavior" by using driving simulators.

Simulation offers a number of advantages, such as repeatability, equality, and universality[83], not achievable in on-the-road studies or if so, only at high cost.

For all the experiments on interaction and communication presented in this work, only systems comprising a single vehicle and one driver have been considered.

13.2 Prototype

For the definition and evaluation of experimental settings in the scope of the research area, prototyping mechanisms within simulated or on-the-road studies were used. A driving simulator allowed an examination of assumptions in a low-cost environment with fixed and precisely controllable preconditions. In general, the results of simulated experiments have to be reviewed by an evaluation based on a real system in the vehicular domain in order to prove evidence and applicability.

On-the-road experiments require a much higher effort for both the system design itself and the processing of the experiment. This is caused, for instance, by (*i*) the prerequisite of integrating sensors and actuators into the car, (*ii*) stability issues of software/hardware, and (*iii*) the necessity for stand-alone operation, which includes auto configuration, error handling and safety functions, etc.

[83] A detailed description of the reasons for simulation is given in the corresponding section on p. 161.

14 Reflecting on the Hypotheses

In this research work two hypotheses with a total of six related subhypotheses have been stated, investigating both the *potential* and the *drawbacks* of implicit data perception and articulation in driver-vehicle environments. Below, each of these hypotheses is indicated and analyzed again, and arguments for accepting and/or rejecting them are presented.

14.1 On Implicit Driver Articulation

A driver's *implicit* feedback toward the car and its assistance systems, respectively, can be applied to support active tasks of vehicle steering or operating in-car devices, and thus has the potential to relieve the driver's level of cognitive load, avoid cognitive overload, and/or not to generate additional distraction.

These assumptions originate from the inference that the driver is not (actively) aware of the information he/she actually delivers.

Subhypothesis I.i The sitting postures of all persons sharing a specific car are significantly diverse, so that that they can be employed for driver identification and accordingly enable personalized car services.

A sitting posture pattern is in some respects an identification feature of an individual, applicable for identifying or authorizing that person in a vehicle. With the investigated method of feature vector calculation and its comparison against the "training set", persons usually sharing a car can be clearly differentiated. The accuracy of the proposed system can be improved when using more significant characteristics as identified during experimental processing.

However, the identification from a sitting posture is not as universal as, for instance, the retina of the eye or the genetic fingerprint. Evaluations of the actual setting and a training set containing 34 data sets showed that the matching for two testing sets vis-a-vis the training database resulted in identification rates of 22.33% for one individual and 31 attempts and 48.08% for the second case with another person and 105 readings. Conducted experiments showed further that the identification rate was prone to changed sitting posture images for an individual to a stronger degree than initially expected.

Nevertheless, and particularly with improved identification algorithms, the recommended system should be applicable in the growing rental car business, for car sharing initiatives, or even in fleet management. With wireless communication technology available everywhere, and the capability of vehicles to join such a network, it would be a great benefit for the driver if his/her personal profile was shared among the different cars he/she uses. This profile would not be limited to covering the simple settings of seat, mirrors, or the radio station, but would also comprise more complex services, such as a distance-based car insurance rate, personalized (subscribed) vehicular services, access to secured internet content, etc.

Subhypothesis I.ii Dynamically captured traces of posture patterns allow driving conditions to predicted prior to their occurrence.

The detection and evaluation of sitting posture patterns, captured together with a multitude of additional sensor data (ECG, video, accelerometer, vehicle-specific data, etc.) during on-the-road studies on a race course, showed that there is in fact a verifiable correlation between a driver's dynamic sitting behavior and specific driving activities.

For instance, and as shown in the Figures 14.1 and 14.2, there is evidence that the leaning direction of the driving person provides information about an ongoing intention of steering (before the corresponding action is initiated)[84].

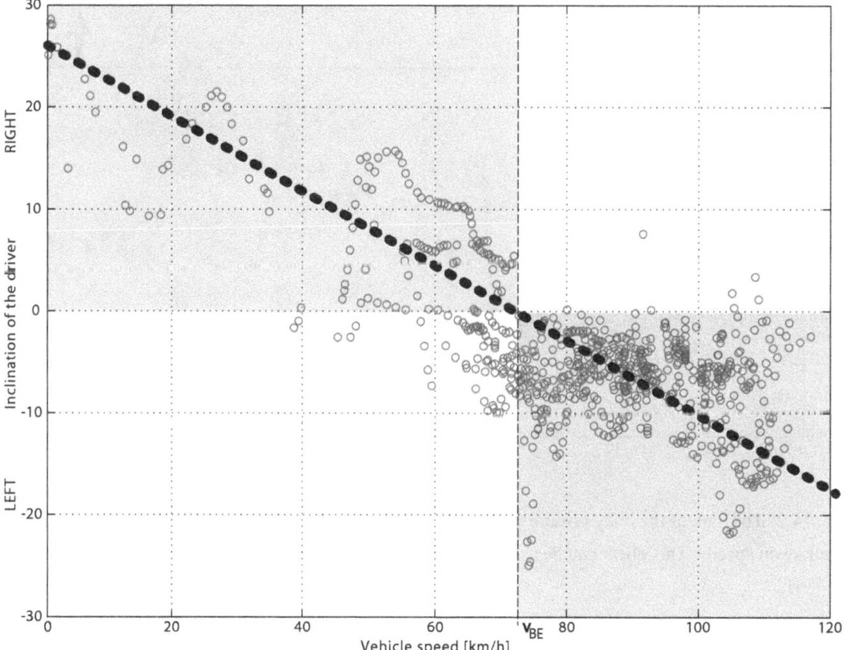

Fig. 14.1: A specific driver's body posture (direction of leaning) correlated to the vehicle speed on a race course driven in counterclockwise direction.

[84] A similar effect was discovered, for instance, by Andreoni *et al.* [244] or Oliver and Pentland [450], [451].

These findings can be applied, for example, in early warning systems operating proactively, e. g., by notifying the driver about activities becoming due shortly and where current sitting posture traces denote no intention of the driving person to perform the corresponding action (and thus allow for the assumption that such activities would be missed).

Of course, an activity recognition system for vehicles operating in real time and with a large number of connected input sensors requires high processing power, generally not available in a car today. But with the steadily rising computation power of CPUs and an ongoing miniaturization of microprocessors, the development of a capable Electronic Control Unit (ECU) should be possible in the near future.

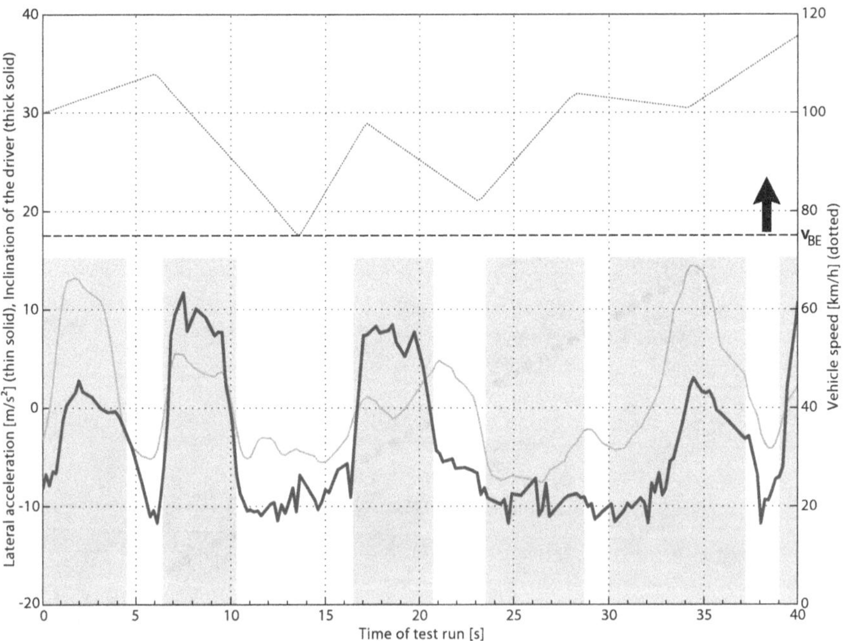

Fig. 14.2: The path of the body posture matches the curve of lateral acceleration (above a driver specific break-even speed). This effect can be used, for instance, for assistance systems operating proactively.

Subhypothesis I.iii Sitting postures can be employed as a means of implicit vehicle control.

This hypothesis is related to Subhypothesis I.ii and uses real-time sitting posture patterns not to infer ongoing activities, but for the systematic execution of control commands. Applications using this additional dimension of input would benefit from enhanced vehicle control capabilities by extracting specific commands directly from posture traces. This approach has the potential to (additionally) relieve the burden on the other utilized sensory channels *vision* and *hearing* (or more generally, the cognitive load of the driver).

As the driver seat is not fully covered by a driver, and, moreover, every driver is sitting in a individual position (as investigated in experiment 4), regions had to be identified that are reachable by any person and in an arbitrary style of driving, before using the seat and backrest as a universal medium for data input.

Three approaches for using the car seat as the universal interaction medium were identified and presented. Associated performance analysis and user acceptance tests for haptic input via the seat or back have not been conducted to date. However, necessary preparations have been completed and user tests may be carried out without any further delay.

14.2 On Vibro-Tactile Driver Notification

Subhypothesis II.i The supplementary use of vibro-tactile feedback in Driver-Vehicle Interfaces relieves the visual and auditory sensory channels of cognitive load, and furthermore reduces the processing time required for the sum of the two motoric sub-processes perception and expression of the Driver-Vehicle Interaction loop.

It has been established that the overall information bandwidth is growing with an increasing number of interaction channels – subject to the precondition that the added channel is appropriate for the required notification or feedback demands. If fulfilled, the additional channel (in this work the sense of touch) would be responsible for taking over a certain amount of the load from the other sensory channels employed.

The sense of touch is particularly suitable for situations where the other two sensory channels (vision, hearing) are restricted by reason of physical limitations, such as glaring or reflecting light, fog, snowfall, day and night vision and changing light conditions (for the visual channel), or motor and traffic noise, passenger communication and cell phone calls, car stereo, etc. (for the auditory notification channel). The tactile channel is claimed to not be affected by these factors.

The different experiments conducted in the scope of this research work showed that the sense of touch holds great potential to support notifications delivered visually or auditorily. Regarding the driver-vehicle feedback loop (see Fig. 1.1 on p. 5), it has been substantiated that the temporal behavior of touch sensations is better compared to stimulations delivered via the visual or auditory sensory channel most of the time. The results of mean response times for the individual feedback channels (as, for instance, presented in Table 12.11) shows significant differences in performance in favor of the sense of touch. On average, response to vibro-tactile notifications is given 14.29% faster than to visual stimuli, and 65.75% faster than to auditory stimulation. However, a conclusion on the hypothesized performance improvement between a system utilizing only vision and hearing and another system employing all of the three modalities vision, hearing and touch cannot be reached.

The performance of haptics compared to vision and/or hearing with regard to the feedback time is very promising and confirms the suitability of delivering important messages via the sense of touch (particularly for time critical tasks or with the aim of reducing the cognitive load of the driving person).

Subhypothesis II.ii The influence of age and gender on stimulus perception and reaction times, which has been proven to affect all the sensory channels vision, hearing, and touch, is the lowest for the haptic sensory modality (and legitimates the extensive application of the sense of touch in future interfaces).

Earlier research studies have provided evidence that the response time in HCI systems is affected by the age of the person(s) involved and it is known that this factor will gain increasing importance in the future due to the advancing age of the population (as highlighted, for instance, on p. 162 or p. 227).

Test persons for the experiments investigating the variances of the three sensory modalities with respect to changes in age or gender had been selected out of a narrow range of age (e. g. a variance of 25.00 ± 5.12 years for the male test participants in the experiment "Simulating Real-Driving Performance") in order to eliminate (or minimize) the influence of age on the experimental results. Moreover, the groups of females and males were analyzed separately to avoid a sex-related impact.

A study on two groups of male persons, divided by the median age, detected only a slight difference of 0.49% between the mean response times of the two groups. Larger differences in age related performance deterioration should be achieved, for instance, when using one group of newly licenced drivers and another one with retired persons. The gender-related evaluation/-analysis indicated that male test persons respond on average 11.13% faster than females.

The variances in the various experiments showed that the margin of deviation for the haptic modality was the lowest for almost all cases of examination. Only the auditory sensory channel for the older test participants in the age-related study had a little less deviation than the sense of touch.

Subhypothesis II.iii Sex and age-related performance deterioration does not affect the order of the sensory modalities with respect to the response time.

The analysis of the dependency of drivers' response times with respect to their age or gender was conducted separately for both parameters. It was discovered that all of the three sensory modalities vision, sound, and touch are highly dependent on these factors.

Evaluations revealed that male test participants respond faster to a stimulus than female test persons at all time. The sequence of the sensory modalities based on the mean response times, with the haptic channel generally ahead of the visual and auditory ones with all 18 test participants and no differentiation of age and gender, remains identical for the two categories (males and females) of test participants and under consideration of the age factor (the latter was evaluated only for the group of male test persons because of insufficient data sets for representative studies on females).

Subhypothesis II.iv The sense of touch as an additional feedback channel in vehicles may replace driver-vehicle notifications delivered by the modalities vision and hearing.

It has been stated and proven in the literature that the more channels are used simultaneously, the lower the load is on an individual modality. However, this is only valid if the additionally used sensory channel is suited to cope with the interaction requirements.

Both experiments showed that the utilization of the sense of touch, side by side with vision and hearing, could enhance any Driver-Vehicle Interaction system and allows for the integration of multimodal interfaces into vehicles, comprising the haptic modality as a third feedback channel, apart from the visual and the auditory notification channels.

15 Experiments: Lessons Learned

While conducting the diverse experiments with tens of test participants, a number of notes and suggestions were collected. The most frequently mentioned comments, which should be taken into account when further improving prototypes and test settings, are:

15.1 Novelty of Haptics

Using the sense of *touch* as a new modality for interaction necessitates instruction or training before comparing it to *visual* or *auditory* interfaces, because it is still uncommon today (particularly in the automotive domain).

15.2 Run-in Experiments

An initial training phase in the simulator studies increased and influenced the later response performance (over all modalities) significantly.

15.3 Annoyances

Notifications based on *vibro-tactile* stimulation are felt to be less disturbing compared to the established channels of notification (*vision*, *hearing*).

15.3.1 Surreal Simulation

A driving simulator performs differently from a real driving journey for several reasons, including

(*i*) a noticeably lower concentration on the task of driving due to the risk-free environment;

(*ii*) ignored traffic rules, such as speed limits (road rage) as therea are no real penalties;

(*iii*) unreal behavior, caused, for instance, by the absence of environmental or engine noise and road vibrations;

15.3.2 Vibration Noise

Depending on the experiment, not only the felt vibration but also the noise caused by vibrating tactors was used as an information source (particularly for the experiments investigating haptic interaction on a more general level).

15.4 Varying Stimulation Parameters

When using vibration parameters, such as frequency or amplitude, for mapping information (e. g. the distance to the next turn) to a particular value, variation of these parameters affected system accuracy. For instance, a vibration of high intensity followed by one with low amplitude (and vice versa) makes it very difficult to estimate the corresponding value in the "real world" compared to less pronounced variation of consecutive notifications.

15.5 Parameter Mapping

Regarding the mapping of e. g. the dimension *distance* to different vibration parameters, test participants reported that *frequency*-based distance estimation is less intuitive and more difficult than a *vibration-amplitude*-based mapping.

16 Conclusion

Finally, it can be concluded that the sense of touch is viable for use in assisting the most frequently used (visual, auditory) sensory channels in the delivery of information from the vehicular systems to the driver or vice versa and thus keeps the workload of a driver low and avoids distraction or cognitive overload.

The experiments conducted in the automotive domain provided evidence that the application of the sense of touch improves reaction performance (or shortens response time) and lessens the workload on the established sensory channels, and in addition helps to improve (*i*) vehicle handling accuracy and/or quality and (*ii*) the safety of both the Driver-2-Vehicle (D2V) and the Vehicle-2-Driver (V2D) interface.

16.1 Applicability

All of the processed experiments in the scope of this research work showed that the sense of touch as an interaction channel performs as well as or better than auditory or visual sensory modalities.

16.2 Additional Workload for Touch Sensations?

In the course of the experiments it was found that the tactile channel creates an additional workload in the initial phase of its application. This is caused by the fact that the sense of touch is relatively uncommon today. However, after a familarization phase, and with later regularity of use, this added workload tended to zero (as for the visual and auditory senses).

16.3 Limitations

A number of restrictions have been identified during the execution of the experiments. All of them have to be considered and/or resolved in order to guarantee safe and accurate system behavior: (*i*) a system comprising the sense of touch has to be fine-tuned and balanced (with regard to parameters like stimulation amplitude, frequency, pulse-pause time, etc.) in order to avoid annoyance or distraction, (*ii*) the sensation of vibro-tactile stimulation is strongly dependent on the individual – the application of a general purpose interaction system is exceptionally difficult or even impossible (depending on the field of utilization), and (*iii*) long-time behavior has not been investigated yet – therefore, it is unclear if the perception tends to blindness towards vibrations on the one hand, or to implicit, intuitive sensations on the other.

17 Future Prospects

17.1 Reconfiguration of Vibro-Tactile Feedback Based on the Driver's Sitting Posture

In this study, the development and evaluation of a hardware prototype of the *vibro-tactile seat* as introduced in experiment 4 on p. 148, with the aim to examine the ability or willingness of a driver to adapt to this user-centered interaction design, is planned. To verify the discussed questions, a real driving experiment should be conducted where the haptic feedback for specific operations, such as a "turn right" operation, will not be fixed to the right edge of the seat, but dynamically adjusted and, for instance, presented in the middle of the seat if a slim person is driving, sitting on the absolute left (as indicated in Figure 12.23, p. 159).

17.2 Reaction Times in Multimodal Interfaces

Apart from the reaction times of individual sensory modalities – as highly stressed in this work – any combination of modalities should generate potential for covering future notification demands. In a large-scale experiment, consequences of combining the different modalities (*vision + hearing*, *vision + touch*, *hearing + touch*, *vision + hearing + touch*) should be investigated with the goal of providing suggestions for the future application of multimodality (particularly in the vehicular domain).

17.3 Integration of Biochemical Features

In this series of experiments, the addtional usage of sensors capturing biochemical features, such as skin conductivity (moisture), respiration, Electrocardiogram (ECG), of the driving person is intended (see, for example, the sections "Biosignal Sensing" starting on p. 204, "Alternatives Supporting the Driver" on p. 84, or "Biometric Identification" on p. 207). The motivation for these studies originates from the proved evidence that any combination of biometric features would enhance a (biometric) identification system.

With the latest technological progress it should now be possible to monitor these features with contactless sensors, invisibly integrated into the vehicle seat (and no longer with generally not accepted devices worn on the body), as, for instance, shown by Aleksandrowicz *et al.* for an ECG device [452], or by Mattmann *et al.* for a respiration sensor [453].

17.4 Additional Sensory Modalities

Some potential can also be seen in the integration of additional sensory modalities, such as using, for instance, the *olfactory* sense to notify about a certain danger (e. g. fire, leak of oil or gasoline) or the *vestibular* sense for providing information on directional cues.

17.5 Theoretical Studies on Haptics and Tactograms

The disposition of the sense of *touch* is still unexplored in some areas and offers potential for improvements in diverse fields of application, not only the automotive domain. Initial studies regarding the applicability of haptics should particularly comprise fundamental research on different vibro-tactile patterns ("*tactograms*") to stimulate specific information with a defined level of attention.

"Technology is a way of organizing the universe
so that man doesn't have to experience it."
Max Frisch (1911 – 1991)
Swiss architect, playwright and novelist

Appendices

A Hardware for Processing Pressure and Vibration

A.1 Pressure Sensing

The utilization of sitting posture patterns to recognize a car driver's identity is a promising, implicit method compared to state-of-the-art identification methods, such as retina scans or genetic fingerprints. Force-sensitive array mats are used for feature acquisition. They are (i) easily integrable into almost any type of seat (the sensor mats are highly flexible and have a thickness of just above $1mm$), (ii) not reliant on the attention of a person using this identification system and furthermore require no active cooperation of this person, and (iii) continuously in operation while the person is seated (sitting in the car is, for instance, an activity lasting for the entire journey). There is evidence, e. g. as stated in a 2003 report of the European Union [454], that people spend an average of $4.5h$ per day sitting in traditional chairs – these findings encourage an adaptation of the posture pattern acquisition system to operate in other application domains as well, such as the office, industry or at home.

A.1.1 Requirements

The pressure-based data acquisition system integrated into the vehicle seat is required to be all-purpose, complying with following general conditions: (i) the sensing area of the pressure mat should be similar to the size of the seat (about $50 \times 50cm^2$), (ii) the force-sensitive mat should be flexible and thin so that it is easily deployable in any type of car and is furthermore not perceived as uncomfortable by the driving person, (iii) the sensing range of the system should cover every driver allowed to steer a car, e. g. persons ranging from 18 to 80 years in age and weighing 40 to $150kg$, and (iv) data acquisition should be precise, calling for a low intersensor distance, paired with a large number of sensors per mat, a high update rate of sensor readings, and a high accuracy of the individual pressure values.

Three systems meeting these requirements have been identified. The decision was made in favor of the "FSA Dual High Resolution System" from Vista Medical[85], rather than the other two systems (from Tekscan and XSENSOR Technologies), mostly because of a local technical consultant versed in the chosen system, and to a lesser degree for technical reasons – as the specifications of all three systems are very similar.

[85] Vista Medical Europe, http://www.vistamedical.nl/?page=products_and_services, last retrieved August 31, 2009

A.2 System Overview

A.2.1 FSA Dual High Resolution System

The chosen system allows the loads to be recorded on a $1.09mm$ thin, flexible sensor mat, built up as a matrix of 32 by 32 piezoresistive sensors. Each of these sensors covers a range from $0kPa$ to $26.67kPa$ (the system specification sheet indicates a pressure range from $0mm_{Hg}$ to $200mm_{Hg}$, however, as neither $Torr$ nor mm_{Hg} are SI units, all values have been conversed into kPa), and overall they encompass an area of $430mm^2 \times 430mm^2$ (initial studies have shown that this size is virtually sufficient for data acquisition and that only exceptionally heavy persons would exceed this area). Each of the $1,024$ sensors covers an area of $7.94mm^2$, and the inter-sensor distance is $5.74mm$ [370]. The system offers a maximum sampling rate of $10Hz$, while typical refresh rates are in the range of $1Hz$ and connection errors sometimes lead to an additional slight delay.

Sensing Software

The original software for pressure sensing as shipped with the system had to be ported to better integrate into the evaluation environment. Particularly in order to adapt to the specific needs of the various experiments, a data collection and evaluation application was implemented on top of this system. A communication with the USB-connected data collection box was established through the OLE[86] components included in the FSA software package, which were extended in order to meet the requirements of the diverse experiments.

Mat Calibration

Prior to its initial use, a calibration of the sensor mat was required for the purpose of compensating for any inevitable constructional heterogenity of individual sensors. During this calibration process, each sensor was matched with a determined weight factor in order to later obtain homogenous, similar pressure values. Repeated calibration would be required every four to ten weeks, depending on the frequency of use.

Measurable Weights

The pressure range ($0mm_{HG} \ldots 200mm_{HG}$ and/or $0N/m^2 \ldots 26,664N/m^2$) detectable with the FSA data acquisition system seems to be quite high. This is true in the case of "normal" sitting, but in some body regions, particularly around the pelvic bones, much higher pressure forces occurs than in the remaining areas. Pelvic bone forces often exceed the maximum sensor value of $26,664N/m^2$ ($200mm_{HG}$), which results in a distortion of results. This type of error is virtually unpredictable because there is no direct connection between pelvic bone force and, for

[86]Object Linking and Embedding is a distributed object system and protocol by Microsoft

instance, the person's weight. Yet a loose forecast of the expected force could be computed from the average pressure applied to the mat.

$$1.0mm_{Hg} = \frac{101,325N}{760m^2} \approx 133.32N/m^2 \quad (A.1)$$

$$p_{max} = 200mm_{Hg} \times 133.32N/m^2 = 26,664N/m^2 \quad (A.2)$$

$$A_{sensed} = 430 \times 430mm^2 = 0.1849m^2 \quad (A.3)$$

$$F_{mat} = p_{max} \times A_{sensed} = 26,664N/m^2 \times 0.1849m^2 = 4,930.17N \quad (A.4)$$

$$m_{mat} = \frac{F}{g} = \frac{4,930.17N}{9.81m/s^2} = 502.56kg \quad (A.5)$$

$$m_{sensor} = \frac{m_{mat}}{nr_{sensors}} = \frac{502.56kg}{1,024} = 0.491kg \quad (A.6)$$

Equations A.1 to A.6 give a rough estimation of the maximum weights measurable by one sensor and/or the entire mat (in case of uniform load). Analysis of a large number of posture patterns has shown that the seat mat is on average covered at approximately 35% – this would indicate that the maximum weight of a person should not exceed $180kg$ ($502.56kg \times 0.35 = 176.25kg$) to ensure correct measurements (and weight estimation). Nevertheless, the weight of a particular test subject could not be exactly determined from the sum of the weights of all the charged sensors due to unbalanced load sharing and *dead space* between the sensor elements.

A.2.2 Other Systems

XSensor X3 Medical Package 4

The pressure sensing system from XSensor Technologies[87] is similar to that from Vistamedical but uses capacitive sensor technology together with a more stable USB hardware interface and more sophisticated software libraries ($16bit$ resolution at a update rate of $100Hz$). That system has been utilized and evaluated for similar experiments, which are not part of this work. The sensor mats of the type "PX100:48.48.02" offer unsurpassed flexibility and durability and consists of $2,304$ sensing points, arranged as a square matrix of 48 by 48 sensors (sensing area of $60.96cm^2 \times 60.96cm^2$, spatial sensing resolution of $1.27cm$).

[87] http://www.xsensor.com/pressure-imaging/, last retrieved August 31, 2009.

Tekscan Dual-BPMS 5315

The pressure sensing system from Tekscan[88] is also connected via USB and operates with sampling rates of up to $10Hz$ at a pressure sensing resolution of $8bit$. The thin-film sensor mats (thickness of only $0.35mm$) are highly conformable and thus optimally suited for operation in body contoured seats. The pressure mats applicable for driver identification studies in vehicle seats are the types 5315 and 5330, both with $2,016$ sensor elements arranged in a matrix of 42 rows (matrix width of $427mm$) and 48 columns (matrix height of $488mm$).

Comparison

Both pressure sensing systems (Tekscan, XSensor) are very similar in spatial resolution and internal construction. These two systems have $2,106$ and $2,304$ sensels, which means their resolution is more than double that of the system from Vistamedical consisting only of $1,024$ sensor elements. Due to this similarity, only one system has been used for comparative studies (the XSensor X3 Medical Package was chosen because of its control interface and the available software libraries); the Tekscan system has not been evaluated in-house.

A.3 Vibration Feedback

For appropriate vibration transmission via the vehicle's seat, a number of conditions and requirements have to be met, e. g. (i) vibrations are required to be strong and punctuated, (ii) devices have to be small-sized, (iii) vibro-tactile actuators need the capability of being integrable into the seat and backrest of the driver seat, and (iv) have to fulfill technical features such as short response times, precise control, and time invariance.

Considering these requirements, only the two technologies (i) piezoelectric actuators and (ii) electromechanical transducers have been identified as suitable (see the comparison of suitable technologies on p. 87). Piezo technology has several disadvantages compared to electromechanical vibration elements: They require high voltage for operation (this presumes a voltage transformer or independent power supply) and provide (very) low vibration amplitudes (these low amplitudes can be partly compensated by using the so-called "sandwich technology"), but on the other hadn they are faster and more precise. Electromechanical vibration elements are disk-shaped, thin, low power consumption devices that provide high vibration amplitudes. Furthermore, both vibration frequency and amplitude are continuously adjustable within a wide range (depending on the type of construction).

After thorough weighing of the pros and cons of both types of systems, the decision was made in favor of the electromechanic technology (comprising voice coils, eccentric mass motors,

[88] http://www.tekscan.com/industrial/, last retrieved August 31, 2009.

etc.), because of (*i*) its well understood operating parameters (frequency, amplitude, pulse-/pause- ratio, etc.), (*ii*) its technical specifications, and (*iii*) the ease of integration into existing systems.

A.3.1 Vibro-Tactile Actuators

Eccentric Mass Motors

Investigation and evaluation of vibro-tactile stimulation was initially done with a custom-built low-cost actuator system. Data communication between host and the vibro-tactile elements was established with an Atmel AVR Mega 32 Microcontroller (RN-Control V1.4[89]) via a connectBlue industrial Bluetooth 2.0 Serial Port Adapter (SPA) OEMSPA441[90]. A ULN2803AG high-current Darlington transistor array, consisting of eight pairs of NPN Darlington transistors, was used for driving eight standard, low-cost Nokia 7210[91] cell phone vibration motors housed in a PVD contactor. For a more detailed technical system description as well as application scenarios see, for instance, Riener and Ferscha [449, p. 240] or Ferscha *et al.* [211].

The controller in its base implementation has the capability to drive eight vibration motors independently; an extension of the circuit with a stage of shift registers would allow enlargement of the vibro-tactile feedback system to control up to 64 vibration elements.

Voice Coil Transducers

For the evaluation of this type of vibro-tactile actuator, high-end linear transducers from Engineering Acoustics, Inc. (EAI[92]) were used. "C-2 Tactors" are miniaturized tactile drivers that can be integrated, for instance, into a seat or mounted within a garment. This kind of tactor is optimized to create a strong localized sensation on the body at its primary resonance (which is in the range between $200Hz$ and $300Hz$, in accordance with the peak sensitivity of Pacinian corpuscles as indicated in "Cutaneous Mechanoreceptors" on p. 89). Multiple tactors can be strategically located on the body and furthermore can be activated individually, sequentially or in groups to convey a specific sensation or provide intuitive "tactile" instructions.

Technically, the "C-2 Tactor" is a linear actuator optimized for usage against the skin. It incorporates a moving contactor that oscillates when an electrical signal is applied, while the skin around is shielded with a passive housing. As opposed to common vibro-tactile stimulators,

[89] AVR Mega 32 μC, 32k ROM, 2k RAM, 1k EEPROM, 32 programmable IO ports, 8 programmable AD ports, http://www.shop.robotikhardware.de/shop/catalog/product_info.php?products_id=10, last retrieved August 31, 2009.

[90] http://www.connectblue.com/products/bluetooth/oem-serial-port-adapter-modules/, last retrieved August 31, 2009.

[91] http://www.yourhandy.de/query.php?cp_sid=26197144e1238&cp_tpl=5503&cp_cat=3571, last retrieved August 31, 2009.

[92] http://www.eaiinfo.com/TactorProducts.htm, last retrieved August 31, 2009.

such as eccentric mass motors, this type of tactor provides a strong, pointlike sensation that is easily felt and localized [455].

The evaluation system was built up of two independent operable and Bluetooth-enabled Advanced Tactor Controllers (ACT) in the version $V2.0$. Each of them can drive eight "C-2 Tactors"; the Ni-MH battery pack ($9.6V$, $2,200mAh$) together with the Bluetooth interface allows hours of mobile operation. Voice coil transducers offer a number of advantages compared to cell phone vibration motors, such as a much higher vibration amplitude (this is quite important when integrating the actuators into the cushion of a seat) and finer adjustment capabilities (vibration frequency is adjustable between $30Hz$ and $340Hz$ in $1Hz$ steps, vibration amplitude can be individually set to one level out of four, and tactor activation/deactivation can be switched within milliseconds).

A.4 Biosignal Sensing

Using additional sensors has the potential to improve a biometric identification system (as indicated in the Subsection "Alternatives Supporting the Driver" on p. 84). In particular, the Electrocardiogram (ECG) is known as a significant characteristic for observing a person's "vital state" which in succession can potentially be applied for vehicle steering or controlling appliances (as proposed, for example, in the section "Improvement Potential" on p. 131).

A.4.1 Electrocardiogram (ECG)

Today, most mobile ECG devices used for measuring heart (or pulse) rate, heart rate variability and other biorhythm related parameters operate with three conductively coupled electrodes (Einthoven ECG), attached to the skin of the person and providing direct resistive contact. Their application in vehicles is almost unfeasable due to the inconvenience and a lack of user friendliness (even the "ultimate" Driver Assistance System, requiring the driving person to attach three electrodes every time he/she gets in the car, would not be accepted).

Aleksandrowicz *et al.* [452] presented a measurement system that obtains an electrocardiogram ECG by using capacitively coupled electrodes. The introduced system is able to measure ECGs through the clothes – without direct skin contact. Although the measurement system is more sensitive to moving artefacts compared to a conventional conductive ECG measurement device and is furthermore strongly dependent on the subject's clothing, it seems useful for at least highly convenient heart rate detection in mobile fields of application. The measurement device additionally avoids skin irritation often induced by the contact gel between skin and the electrodes.

The proposed capacitive measurement system could be, for example, integrated into a vehicle seat with two electrodes embedded into the back, and the reference electrode integrated into the seat. This system would then operate fully autonomously and attention-free, and thus would be another building block for the class of implicit operating sensing systems.

In the research studies conducted in the scope of this work, a common body-mounted ECG device "HeartMan 301" from HeartBalance AG was used. This appliance is small-sized, light-weight, operates reliably, records up to 24 hours with one battery pack, and delivers highly precise data in real time. Data sets are either transmitted wirelessly via a Bluetooth communication interface or stored in European Data Format (EDF)[93] on an integrated SD memory card.

A.4.2 Additional Sensors

Other types of sensors, such as a respiration or Galvanic Skin Response (GSR) sensor, could be integrated into specific control elements of the car (e. g. the steering wheel, gear lever knob, handbrake, etc.) for implicitly recording bio-chemical signals representing factors like cognitive load, mental state, and stress level of the driver. Collecting these additional data has the potential to accelerate the development of a more precise model of human (or driver) behavior, which in turn should lead to an increased reliability and/or accuracy of the D2V communication interface.

With the exception of ECG devices, no other types of sensors have been used in the scope of this work for capturing bio-chemical signals.

A.5 Environmental Sensing

Mobile internet and wireless communication technologies as well as the Global Positioning System (GPS) are present in vehicles today, enabling the observation of the static environment (e. g. information on Points of Interest POI, on road conditions or toll information, or on road signs) and additionally allow analysis of dynamic traffic relevant data, such as weather conditions, road traffic density or traffic jam signals, petrol price information, etc. in real time. All of these attributes can help to improve vehicle handling by automatically adapting affected Driver Assistance Systems without burdening the driver with active notification or cooperation. Examples would be, for instance, (*i*) automatic re-routing to prefer or avoid toll roads according to a predefined driver-dependent policy, (*ii*) decreasing the vehicle speed in response to traffic jam warnings, or (*iii*) automatically adding an intermediate stop on the route plan at a gas station with very low fuel prices, with the petrol tank filling level reporting some capacity left.

[93]http://www.edfplus.info/, last retrieved August 31, 2009.

A.6 Vehicle-Specific Data

The standardized On-Board Diagnostics (OBD) interface allows data acquisition for approximately fifty vehicle-dependent parameters, such as velocity, motor RPM, throttle position, different air and gas pressure values, motor temperatures, etc.

The analysis of these parameters should help to better estimate the current state of the vehicle in real time, however, is limited by the fact that the OBD interface operates serially. This means the more parameters requested, the lower the update rate of each sensor reading. A update rate between $5sec.$ and $6sec.$, which has been indicated in a prototype collecting less than 10 attributes, would be generally unacceptable for interacting with a vehicle's systems or applications in real time.

To solve this issue, two alternatives have been identified, namely (i) integrating an external measurement system (a dedicated data logger) into the car and connecting it to all desired sensors (this option has been evaluated with a "DL2" high end professional GPS data logger by RaceTechnology Ltd.[94]) or (ii) establishing a direct connection to a sensor using available vehicle buses (e. g. CAN, MOST, FlexRay). For the second option the implementation and integration of a new ECU with real-time access to the bus(es) (and interconnected sensor devices) would be required.

[94]http://www.race-technology.com/dl2_2_28.html, last retrieved August 31, 2009.

B Biometric Identification

In this section, different biometric identification technologies are introduced and each compared with the method *"Identification from Sitting Posture Patterns"*. Furthermore, requirements for an in-car usage of the different approaches are stated and related terms (authentication, open-set versus closed-set, etc.) are defined.

Applicability

Mobile internet services have started to pervade into vehicles, approaching a new generation of networked, "smart" cars. With the evolution of in-car services, particularly with the emergence of services that are personalized for an individual driver (like road pricing, maintenance, cause-based taxation, insurance and entertainment services, etc.; for further applications see Fig. 4.1), the need for reliable in-car identification (authentication) technology has arisen.

B.1 Related Terms and Definitions

Woodward *et al.* [456] defined the term *biometric* concisely as "[..] the automatic recognition of a person using distinguishing traits". A definition of *biometrics* in more detail would be, for instance, "[..] any automatically measurable, robust and distinctive physical characteristic or personal trait that can be used to identify an individual or verify the claimed identity of an individual" [456].

B.1.1 Human Behavior

The behavior of an individual in dynamic environments is dependent on numerous dimensions, such as context and time. Bobick [457] divides the expression "behavior" into three subterms (i) movements (these are the basic behaviors and they have no linkage to situational context or temporal structure), (ii) activities (these are short sequences of movements combined with some temporal structure), and (iii) actions (these are recognized from activities, and interpreted within the larger contexts of participants and environments).

B.1.2 Identification and Verification (Authentication)

Biometric identification refers to identifying a person on the basis of his/her physiological characteristics and biometric identifiers. Identification and verification (or authentication) are significantly different terms: Identification in a biometric system tries to clarify the question "who a specific person is" (this is a 1:N relationship, comparing the currently acquired pattern against biometric profiles from a database), while authentication attempts to answer the question "is this person A?", after the user claims to be A [456, p. 2] (the latter is much more complex than the first, however, could be potentially solved by using public/private key ciphering techniques).

Authentication may be defined as "providing the right person with the right privileges the right access at the right time". In general, the security community distinguishes among three types of authentication [456, p. 6], [16, p. 27] (ordered from least to highest security and convenience).

(*i*) Physical ("something you have"), e. g. an access card, smart card, token, or key.

(*ii*) Mental/Cerebration ("something you know"), for instance a password, PIN code, or piece of personal information (the birthday date of one's mother or the name of one's dog).

(*iii*) Personal ("something you are"), which is a biometric characteristic dividable into an active (explicit) type, such as the interpretation of retina/iris, voice, face, or fingerprint, and a passive (implicit) type, comprising, for instance, sitting posture pattern evaluation.

The third class "Personal" is the most secure verification option because it cannot be lost, stolen or borrowed (such as keys), cannot be forgotten or guessed by an imposter (as, for example, a password), and forging of a biometric characteristic is even impossible. In contrast, physical or mental systems have the disadvantage that they are not based on any inherent attributes of a person.

Implicit authentication methods based on biometric features would prevent the driver from additional cognitive load and distraction, which is particularly important in situations where the person is continuously highly challenged, such as, for instance, during a city trip at rush hour.

B.2 Person Identification (Authentication) in Vehicles

Services that demand unambiguous and unmistakable (continuous) identification of the driver have recently attracted many research efforts, mostly proposing video-based face or pose recognition techniques or acoustic analysis (e. g. voice recognition). An overview of the characteristics of appropriate biometric identification technologies for vehicular use is presented in Fig. B.1; a description of biometric security technologies in more detail can be found, e. g., in Liu and Silverman [16]. It is assumed that all of the explicit identification techniques can be replaced by implicit driver identification methods such as sitting posture evaluation (at least for selected secondary vehicular applications).

Apart from biometric identification technology, sensor signals captured from acceleration or brake pedals, OBD interface (e. g. vehicle speed, engine RPM), steering wheel, etc., and

Characteristic		Biometric Identification Techniques (suitable for in-car usage)							
		FINGER-PRINT	HAND GEOMETRY	RETINA	IRIS	FACE	SIGNATURE	VOICE	SITTING POSTURE
	EASE OF USE	High	High	Low	Medium	Medium	High	High	**High**
	ERROR INCIDENCE	Dryness, dirt, age	Hand injury, age	Glasses	Poor lighting	Light, age, glasses, hair	Changing signatures	Noise, colds, weather	**Age, weight**
	ACCURACY	High	High	Very high	Very high	High	High	High	**Medium**
	COST	*)	*)	*)	*)	*)	*)	*)	***)**
	USER ACCEPTANCE	Medium	Medium	Medium	Medium	Medium	Veryh high	High	**Very high**
	REQU. LEVEL OF SECURITY	High	Medium	High	Very high	Medium	Medium	Medium	**Medium**
	LONG-TERM STABILITY	High	Medium	High	High	Medium	Medium	Medium	**Medium**
	IMPLICIT ACQUISITION	No	No	No	No	No	No	Yes	

*) *The large number of factors involved makes a simple cost comparison impractical.*

Fig. B.1: Comparison of biometrics (adapted from Liu *et al.* [16]).

associated with the driver's behavior, should provide substantial benefit for personalized in-car applications. In [53], Erzin *et al.* reported on driver identification experiments utilizing driving behavior (particularly speed variants and pedal pressures) and found that driver identification is not possible on these parameters, however, that the "vehicle context" allows verification and reaction to the driver's physical or mental condition (alert, sleepy, or drunk) to a certain degree.

None of the known research work (including that of Erzin *et al.* [53]) actually utilized implicit identification capabilities (static or dynamic) collected from the driver's body characteristics, such as the sitting posture pattern analysis method proposed here.

For utilizing personalization specialized to the vehicular domain, the identification demand can be divided into two groups, (*i*) personalized settings, such as seat and mirror adjustments, radio station presets, calibration of the running gear, chassis set-up, horsepower regulations, etc., where the adaptation to predefined settings requires an identification once at the time of boarding, and (*ii*) personalized services, such as a car insurance rate, road pricing, cause-based CO_2 taxation, etc., demanding a continuous identification of the driver while seated (and operating the car).

The latter class, combined with additional sensors and/or processing capability, would allow for new, advanced fields of vehicular applications, such as (*i*) authorization (increased vehicle safety, e. g. by determining if the person sitting in the driver seat is authorized to steer the vehicle), or (*ii*) safety (promoting safe driving by monitoring a driver's behavior, e. g. to determine if the driver is sleepy or drunk).

B.3 The Identification Process

Identification is referred to as selecting the correct identity of an unknown person from a database of registered identities (as depicted in Fig. B.2). This is a *1:N* (a *one to many*) matching process – the system is asked to complete a request by comparing the person's (driver's) biometrics against all the biometric templates stored in a database. The system can take either the *best match* or it can score the possible matches and rank them in order of similarity [456, p. 2].

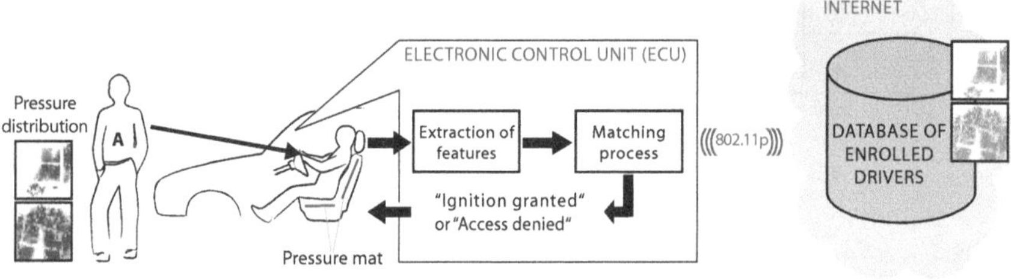

Fig. B.2: Driver identification from sitting postures in vehicles (implementation as a universal, distributed pattern recognition system).

B.3.1 Open-Set and Closed-Set Person Identification

Biometric identification can be classified into *closed-set* and *open-set* detection [53]. In both cases the unknown person requests access to a service without an explicit identity claim.

Open Set

In this case, a reject scenario is included – the system identifies the person only if there is an exact match to one of the N registered persons and rejects him/her otherwise. Hence, the problem becomes an $N+1$ class identification problem including one reject class.

Closed Set

The individual is (always) classified as one of N registered persons, namely that with the most similar biometric feature in the database. Access is granted with the personalized profile of the best match while a reject scenario is not defined (and thus, imposters are not handled).

Summary

The *open-set* identification approach is well suited for a vehicle access application in which several drivers might be authorized to access a single car (if a specific driver is granted access with the profile of another person, nothing hazardous would happen).

On the other hand, in the authentication scenario where a driver's condition is deduced from the monitored driving behavior in order to assure safe driving or react appropriately on situations where the driving person is tired or drunken, for instance, the *closed-set* identification technique would be required to ensure system safety and stability (otherwise, the mental and/or physical condition of a driver would be potentially matched against that of another, false person).

B.3.2 False Accept and Reject Rates

Depending on the identification goal, either a *positive* or *negative* identification strategy would be better suited. The *positive* identification technique tends to determine if a given person is already in the database. Such a method is applied, for instance, when the goal is to prevent multiple uses of a single identity. The *negative* identification approach determines if a given person is not in a "watchlist" database. Such a method is applied, for example, when the goal is to identify persons registered under several identities.

For a rating on the accuracy of a biometric system, the two methods False Accept Rate (FAR) and False Reject Rate (FRR) are widely used. Both methods focus on the system's ability to allow limited entry to authorized users. FAR and FRR are interdependent and can vary significantly [16, p. 32]; thus, both measures should be given together (e. g. plotted one against the other). FAR (see equation B.1) gives the percentage of users who claimed to be a specific identity and that the system has falsely accepted for the claimed identity. On the other hand, FRR (equation B.2) is the percentage of users who claimed a certain identity for which the system has either false rejected them or does not have the capability to acquire their biometrics.

The terms False Accept Rate (FAR) and False Reject Rate (FRR) are defined as

$$FAR = \frac{\text{No. of false accepts}}{N_a + N_r} \tag{B.1}$$

$$FRR = \frac{\text{No. of false rejects}}{N_a} \tag{B.2}$$

B.4 Ideal Biometric Identification Systems

The significance of any biometric identification system is determined by the assortment of appropriate features. The studies and experiments on driver identification or activity recognition from sitting posture patterns in this research work are aligned with the recommendations for ideal biometric systems presented by Jain *et al.* [404, p. 92], which are (*i*) universality, (*ii*) collectability, (*iii*) uniqueness, and (*iv*) permanency. For details concerning these biometric characteristics see the section "Identification and Authorization" starting on p. 121 or the corresponding articles [426], [405].

Jain *et al.* additionally claimed that these four requirements are not always feasible for useful biometric systems and that other issues have to be considered in the realization of an authentic system.

Performance

This attribute is, apart from properties like wastage of resources (e. g. energy consumption), etc. characterized by the two characteristics False Accept Rate (FAR) and False Reject Rate (FRR) [458, p. 57]. In an optimal biometric system both of these error rates tend to zero – the biometric system is then indicated as "optimal".

For an identification system operating in vehicles the characteristic *performance* is additionally influenced by the factors (*i*) real-time capability and (*ii*) available computation power of the in-car systems.

Acceptability

This criterion denounces the willingness of people to accept the usage of a particular biometric identifier (in everyday life).

In the applications and/or experimental settings used in the scope of this work, identification was done with force sensitive mats. For that reason the thin sensor arrays were invisibly integrated into seat and backrest of the driver's seat, thus this criterion should be universally fulfilled and can be neglected henceforth.

Circumvention

This term refers to an estimation of how easy it is to fool the system through fraudulent methods. The system architecture recommended for the implicit data acquisition system based on sitting postures should ensure system safety for the following reasons: (*i*) the sensing hardware is a closed system integrated into the seat, (*ii*) the data processing unit (ECU) is embedded into the car and both obstructed and inaccessible after installation, and (*iii*) the data transfer over the vehicular bus systems itself will be made encoded.

To fool the identification system forging of a person would be required, meaning that the imposter should have the same physical shape of back and bottom, a similar weight, and a equal sitting behavior as the claimed identity – particularly the last feature should be practically impossible to reproduce.

B.5 Potential and Limitations

The identification of persons and/or the recognition of activities in the highly dynamic automotive environment based on an evaluation of sitting postures would offer great potential on the one hand, however, is subject to several issues on the other. For instance, sudden movements would cause erroneous data collection, resulting in an increased False Accept Rate and/or False Reject Rate and in succession leading to unexpected system behavior. Calculating extended features from the posture patterns or using improved evaluation techniques (such as those established for face recognition in still or moving images) can help to enhance the reliability and stability of the system.

Further improvements should be achieved when combining several biometric identification techniques, such as additionally incorporating (*i*) ECG traces contactlessly collected from electrodes in the safety belt or the seat (for details see "Electrocardiogram (ECG)" on p. 204) or (*ii*) the driver's skin conductance measured with the help of a Galvanic Skin Response (GSR) sensor[95, 96] embedded into the steering wheel or gear shift.

Biometric-based implicit identification of an individual would enable a new class of vehicular services, such as a safety function that authorizes a person automatically when seated on the driver seat and would only allow permitted persons to start or drive the car (this accreditation would be denied to all others). Posture pattern based identification technology could furthermore be conceived as an effective car-theft protection system.

However, on the other hand, notable limitations have been identified. Suppose for instance, that the driver of a car gets injured or incapacitated and someone else is asked to drive him/her to a hospital – would there be any chance to override the authorization process, allowing the helping person to use the vehicle in a "rescue mode" without a personalized profile?

---

[95] Merriam-Webster's Medical Dictionary, http://medical.merriam-webster.com/medical/galvanic+skin+response, last retrieved August 31, 2009.

[96] Biosignal Analysis at University of Kuopio, http://bsamig.uku.fi/research.shtml, last retrieved August 31, 2009.

C Physiological Senses and Proprioception

Apart from the *traditional* or *exteroceptive* senses (vision, hearing, touch, smell, and taste) physiologists define a second group of sensory modalities called *physiological* or *interoceptive* senses.

Proprioception is another self-contained and distinct sensory modality providing feedback about the internal status of the body which indicates position (of the different parts of the body in relation to each other) and movement of the body[97].

The five traditional senses, first described by the Greek philosopher Aristotle, are well-established and their applicability for vehicular applications has already been shown. Particularly the senses *vision*, *hearing*, and *touch* are extensively stressed in this research work, e. g. in the section "Excursus: Sensory Modalities" starting on p. 41.

C.1 Physiological Senses

Due to a steadily rising volume of information and the associated increase of load of the classical information channels, the *psychological* senses continue to gain in significance. These senses, as indicated in Fig. C.1, have not been studied in detail to date and are rarely unemployed in most fields of application.

Physiological Senses			
SENSORY MODALITY	PERCEPTION	ORGAN	IN-CAR DEVICE OR APPLICATION
Equilibrioception (Vestibular sense)	Balance, Gravity	Organ of Equilibrium (Inner ear)	Navigation system, Route guiding Directional information
Thermoception	Temperature (Heat)	Skin	Air conditioning system Engine/Fuel/Coolant water
Nociception	Pain	Tissues, Skin, Joints	Adjustment of chassis, Comfort level of the car seat

Fig. C.1: Physiological senses and corresponding fields of application in vehicles (adapted from Silbernagel [17]).

Although the current work is focused on the utilization of the *traditional* senses (particularly the sense of *touch*) this subsection is dedicated to a short overview of the *physiological* senses with the aim to identify potential for their future application in vehicular systems or services.

[97] The fragmentation into *proprioception*, *interoception*, and *exteroception* was first elaborated by Sherrington, 1906 [459].

C.1.1 Balance

Equilibrioception or the *vestibular* sense is sensitive to both *balance* and *acceleration force* (or *gravity*), and thus can be imagined as an additional sensory channel for navigational tasks or route guiding with systematic altering of the sensory channel. Although the vestibular sensory system has not yet been well researched, it already has attracted an interest in virtual reality applications such as flight simulators or race car games – an adaptation and evaluation of the sense of *balance* for real-world applications would be the next logical step.

The vestibular organ is situated in the inner ear and is mostly related to cavities containing fluid (equilibrioception is normally sensed by the detection of acceleration occuring in the vestibular system). *Balance* is affected by other senses, such as the visual sense (there is evidence that it is harder to stand on one foot with eyes closed than with the eyes open).

Latest research work confirms the significance of the *vestibular* sensory organ for future control tasks: Scinicariello *et al.* [460] reported that Galvanic Vestibular Stimulation (GVS) could be used for systematically controlling an individual's posture (potential for in-car use: avoid fatigue, improve vehicle handling in cornering stuations, etc.). Wardman *et al.* [461] found that GVS response diminishes when visual or tactile information is available and increases when proprioceptive information is limited or test subjects are placed on an unstable surface (e. g. standing on foam). Furthermore, they identified that GVS response is highly automatic and does not become habitual (even when self administered repeatedly). Furman *et al.* [462] investigated the role of attention in sensory-motor processing of vestibular and combined visual-vestibular information during seated rotations using a dual task interference approach. They found that (*i*) vestibular stimulation has an influence on information processing tasks, (*ii*) older subjects had longer reaction times for all combinations of stimulus condition and reaction-time tasks compared to younger subjects (additionally, reaction time decreases as the task complexity increases), and (*iii*) vestibular sensory input can inhibit another sensory modality (auditory).

In [463], Bacsi and Colebatch confirmed that the vestibular sense plays a significant role in maintaining postures, particularly under conditions in which other sources are diminished or absent. MacDougall *et al.* [464] studied the effect of a pseudorandom binaural bipolar Galvanic Vestibular Stimulation (GVS). In general, GVS was well tolerated by test participants and their results suggest the use of GVS for modeling postural instability of vestibular origin both quantitatively and qualitatively. Laschi *et al.* [465] developed a vestibular interface for controlling robotic systems based on the detection of a human's motion intention (instead of the movement itself). They validated and confirmed the hypothesis that head movements can be used to detect a person's intention to execute a steering action during locomotion (slightly in advance) and that this signal can be used to control navigation tasks of a robotic system.

Maeda *et al.* [466] presented a novel interface based on Galvanic Vestibular Stimulation to support humans in GPS based walking navigation, for collision avoidance or in pedestrian flow control by altering the balance of test participants. With minor customization of the control interface, in-vehicle operation for control tasks, navigation or path finding should be possible.

Aside from the related work described above, the significance of using the *vestibular* sensory organ for (future) control tasks has been impressively shown in a vast number of articles published during the last three years, e. g. by Bacsi *et al.* [463], Taube *et al.* [467], MacDougall *et al.* [464], Sandler and McLain [468], Amemiya *et al.* [469], Aoki *et al.* [470], Harris *et al.* [471], Nagaya *et al.* [472], Laschi *et al.* [465], Jenkin *et al.* [473], Golding *et al.* [474], Amemiya *et al.* [232], or Carr and Newman [475].

Applicability

The *vestibular* sensory system has only been of interest for a short time, but has meanwhile been successfully applied as an additional interaction channel in numerous virtual reality simulators, and thus should also offer potential for being used in "real" vehicular applications. With the increasing load on the traditional senses (shown in table C.1), it is necessary to consider new forms of interaction or communication. *Equilibrioception* has the potential to cover these requirements, in particular assistance systems for (*i*) route guiding ("vestibular provoked navigation") or (*ii*) providing directional information in general would be imaginable applications.

C.1.2 Temperature

Thermoception (or *thermoreception*) is carried out by the skin and is the sense of perceiving temperature, where *heat* conforms to temperatures above the body temperature (maximum responsiveness at 45^oC) and *cold* corresponds to temperatures below the body temperature (highest sensitivity at 27^oC). Cooling or cold receptors occur in the skin about 10 to 30 times more frequently than heat receptors, clearly operate faster, and subjects are more sensitive to cold than to warm stimuli [196, p. 313]. The density of both types of receptors differs depending on the body region (highest density for both types is on the nose, lowest density is on the upper arm) and they do not overlap. Furthermore, thermoreceptors are more responsive to a change in temperature than to a constant temperature [476].

Applicability

The investigation of how these receptors work in detail is still in progress and a direct employment in motor vehicles cannot be deduced at this time.

C.1.3 Pain

Nociception or *physiological pain* is the unconscious, unpleasant perception of near-damage or damage to tissue [196, p. 316]. Nociceptors[98] are sensory neurons that are found in any area of the body (skin, muscles, bones, and tissues) and are sensitive to mechanical (pressure), thermal (hot, cold), or chemical sensations [476]. This type of receptor needs a strong stimulation for an excitation and, contrary to other sensory modalities (including thermoception), does not adapt – a continuous application would not lead to a reduction of excitability.

Applicability

The sensation of pain is always a subjective experience – a utilization would therefore always be associated with problems regarding personal sensitivity to pain (*algesthesia*). Pain receptors are known to be highly important for avoiding accidents [476] – this characteristic may offer potential for an application in the automotive domain (e. g. in a crash or traffic jam warning system or an automatic adjustment of the chassis depending on the road bed).

C.2 Proprioception

There is some confusion in the literature regarding the relationship of the terms *haptics*, *tactile*, *touch*, *cutaneous*, etc. and particularly between *proprioception* and *kinesthesis*.

For instance, the term *haptics* is often used equivalent to *somesthesis*, which is (e. g. in the opinion of R. Cholewiak [18]) false. Recently, *haptics* has become synonymous with all tactile experience and this is also an incorrect usage.

In this part of the appendix, these topics are being explored from different points of view with the aim to clarify their interrelationship.

C.2.1 Requests for Clarification

To get a clear understanding of the relationship between the terms surrounding *haptics*, a number of experts (scientists, psychologists) in this research area have been consulted. Their original responses to questions related to the difference between *kinesthesis* and *proprioception*, as well as the interrelationship of related terms are given in the following. A summary of their views, finally leading to a definition of the terms for the utilization within this work, is given in the sectionh "Definition of Terms" on p. 221.

[98] International Association for the Study of Pain (ISAP), http://www.iasp-pain.org//AM/Template.cfm?Section=Home, last retrieved August 31, 2009.

Gabriel Robles-de-la-Torre, International Society of Haptics, [477]

"Dear Andreas,

The discrepancies in the meanings of "proprioception" and "kinesthesia" reflect the complexity and multiple sources of information contributing to these phenomena (cutaneous, muscular and joint receptors, etc), and also the fact that these phenomena have not been as extensively researched as vision, for example. As a result of this, much remains to be understood and clarified. For example, until recently it was thought that limb motion sensing relied on muscle and joint receptors, but now it is known that cutaneous information also contributes to this capability. So there is a lot still to be learned. As the phenomena become better understood, the definitions will, hopefully, be less ambiguous.

I think what matters at this stage is to carefully clarify what you mean when using the terms, so that you minimize the ambiguity of the definitions. In a generic sense, "kinesthesia" and "proprioception" can be used interchangeably (both including motion and position of limbs). But, for example, when specifically referring to motion or position sense in some joints, it is better to be more specific. For example, it is known that there is a sense of movement but not a static position sense in a joint in the index finger [478]. When using "proprioception" to refer to this joint's capabilities, it would be necessary to clarify that this term implies movement sense and does not include position sense.

I hope this helps.

Gabriel"

Susan Lederman, Department of Psychology, Queen's University, Ontario, Canada, [479]

"Hi Andreas, alas you are certainly not the first to be confused! My take on this is that the sense of touch consists of two subsystems: cutaneous (involving the skin and associated receptors) and the haptics subsystem (consisting of combined inputs from both cutaneous AND kinesthetic (includes receptors in muscles, tendons and joints) subsystems.

There has been, and continues to be, considerable confusion about the terms proprioception and kinesthetic. The former term, as you say, was first used a long time ago by Sherrington and by many other neurophysiologists after that. They have described proprioception as the broader term, including both the kinesthetic subsystem, which provides our sense of where our limbs are in space and how they are moving, AND the vestibular system in the inner ear, which is responsible for our sense of balance and coordination. It is true and very confusing that the two are still sometimes used interchangeably, although this continues to decrease with time. For those of us studying the sense of touch per se, most no longer use the term proprioception (i. e., we regularly exclude the vestibular system) in our discussions, and thus only focus on the role of the kinesthetic system."

Roberta Klatzky, Department of Psychology, Carnegie Mellon University, Pittsburgh, Pennsylvania, [480]

"Excerpt from Wolfe *et al.* [481, Chapter 12]: By its most narrow definition, the term touch is used to refer to the sensations caused by mechanical displacements of the skin. These displacements occur when you are poked by your 4-year-old nephew, licked by your dog, or kissed by your significant other. They occur any time you grasp, wield, or otherwise make contact with an object. We will use the term tactile (the adjective form of touch) to refer to these mechanical interactions and will expand the definition of touch to include the perception of temperature changes (thermal sensation), the sensation of pain, which occurs when our body tissues are damaged (or potentially damaged) in some way, itchiness, and the internal sensations that inform us of the positions and movements of our limbs in space. Collectively, these internal sensations are known as kinesthesis when they arise from muscles, tendons, and joints, and they are part of a broader system known as proprioception, which includes the vestibular system as well (see [481, Chapter 15]). The technical term for all these senses put together is somatosensation."

Vincent Hayward, Institut des Systemes Intelligents et de Robotique, Université Pierre et Marie Curie, Paris, France, [482]

"Hi,

I am just as confused as you are. And even physiologists of great authority may not be consistent with each other or with themselves. It also depends on if you take the historical perspective or today's practice, if you talk to a neurophysiologist or a cognitive scientist. I am a mere engineer and also like to call things always the same way, but I guess one has to accept the fuzz associated with studying living creatures.

Haptic, I just found out, was invented with Géza Révész (Hungarian-Dutch) last century. I understand his idea was to find a word for the perception of objects through touch like you have auditory or visual perception.

The approach I have taken is simply to take those words at their etymological value, this way it's more clear to me.

Prorioception: "proprio" + "ception" awareness of your own body. Kinesthesia: "Kine" + "thesia" awareness of movement (normally one includes effort even if there is no net movement but there is internally in the muscles). So you could put proprioception "above" kinesthesia since you could think that proprioception is more perceptual whereas kinesthesia refers more to raw sensations. It's very confusing since it's known that the skin "cutaneously" gives proprioceptive information (in the joints), which on second thought makes a lot of sense, although no "kinesthetic" receptors are involved. There are lots of such cross connections.

Haptics puts all this together into a single package (like in vision there is B&W vision and color vision which are quite separate but once it reaches your consciousness there only is one

world out there). If you manipulate a ball, it's a ball not a piece of cutaneous input coming from the hands a bit coming from the joints, signals converging to the muscles, others returning back up from the tendons not counting the efference copy sending signals to the cerebellum (and to the basal ganglia); a huge amount of stuff going on... all this results in just a ball. So you call that haptics. I suppose Révész had a good idea! This way I don't lose sleep on it.
Hope that helps,
Vincent"

Hong Z. Tan, School of Electrical and Computer Engineering, Purdue University, West Lafayette, Indiana, [483]

"[..] You are quite right to point out the subtle differences between the terms proprioception and kinesthesis, but as far as I am aware, most of us use the two terms interchangeably. For whatever reasons, people tend to use 'tactile' and 'kinesthetic' these days as opposed to 'cutaneous' or 'proprioception'. I do not think that you need to fuss over the subtle differences unless it is part of your research.
For references, you may also want to read the following two chapters: Clark and Horch, [484] or Loomis *et al.* [485].
Hong Tan"

Roger Cholewiak, Cutaneous Communication Laboratory, Princeton University, New Jersey, USA, [18]

"Hello Andreas,
Unfortunately, your confusion is common, partially owing to the adoption of some terms by the engineering community in a way that goes contrary to to historical sensory and perceptual usage. The most egregious example is the use of haptics to be equivalent to somesthesis, as you describe in the first line of your 3rd paragraph. Haptics, all the way back to Boring and Gibson's usage, involves a combination of touch and kinesthesis. That is, haptics always involves active touch - exploratory or manipulative behaviors. The kinds of activities that we conducted in our laboratory for many years, involving passive perception of tactile stimuli, is tactile or cutaneous sensitivity, or touch, NOT haptics. The Verrillo/ Geldard/ Craig/ Sherrick/ Gescheider/ von Bekesy bodies of literature, as well as mine, all involve only cutaneous sensitivity and pattern perception, not haptic perception. Recent work by Lederman/ Klatsky/ Tan/ Weisenberger involves haptics. Recently, however, haptics has become synonymous with all tactile experience. And that is an incorrect usage.

Regarding whole body experiences, proprioception refers to both vestibular and kinesthetic sensibility. That is, proprioception is our awareness of our body's position and movement in the environment – providing feedback of our body's movement in space. Kinesthesia is reserved for our perception of limb (joint) positions and limb movement in 3D space – through receptors in muscles/ tendons/ joints, while the vestibular senses involve the semicircular and otolith organs, sensing head/ whole body acceleration of one type or another – linear or rotatory movement. Consequently, I agree that somesthesis involves both passive and active tactile experiences and exploration – the first involving only the cutaneous senses, the second involving touch and kinesthesis, so:

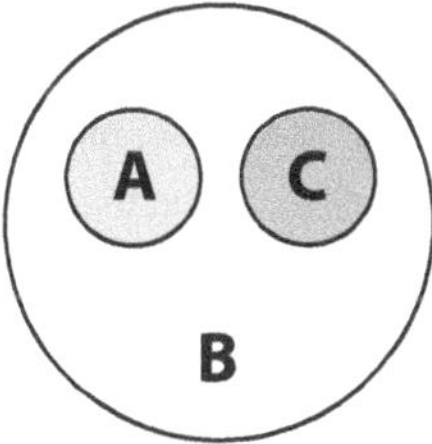

Fig. C.2: Interrelationship of kinesthesia (A), proprioception (B), and the vestibular sense (C) (taken from R. Cholewiak [18]).

I hope these comments help to understand how I view these terms. That said, my view does not agree with usage in some of the literature, just because I feel people have become loose with their definitions. That means, to me, that the meanings of the words are becoming lost...
With best regards,
Roger Cholewiak"

C.2.2 Definition of Terms

The different utilization of the terms related to the sense of touch (as confirmed by the experts) and the associated discrepancies reflect the complexity of these sensory modalities as well as the fact that touch has not been researched as extensively as, for instance, visual or auditory channels.

Haptic

The term *haptic* (Greek, "to lay hold of") was firstly used (and defined) by Géza Révész (1878–1955), a Hungarian-Dutch psychologist[99], as "a word for the perception through touch following the understanding of auditory or visual perception" (see Pieron [486], Busato [487], and Hayward [482]). Schiffman [398, p. 370] defines *haptic* as a "sensory-perceptual channel that refers to the combined input from the skin and joints". Considering the definition and/or usage

[99] CHC61 IEEE Competition: Haptic History (Universidad de las Ciencias Informaticas), http://hapticshistory.chc61.uci.cu/haptic/site/pages/Geza_Revesz.php, last retrieved August 31, 2009

of haptics by Boring [488] and/or Gibson [489], haptics involves a combination of touch and kinesthesis (this implies that haptics always involves active touch – exploratory or manipulative behaviors).

As a consequence, the application of *haptics* as a general term for all touch-related notifications – as often used to date – is no longer valid.

Proprioception, Kinesthesis

Since Sherrington's work (1906) [489, p. 55–56], [490], the sensory organs are subclassified into (*i*) exteroceptors (receptors in eyes, ears, nose, mouth and skin), (*ii*) proprioceptors (sensory organs in muscles and joints as well as in the inner ear (=vestibular organ)), and (*iii*) interoceptors (various nerve endings in the internal organs), where exteroceptors are responsible for information (stimulation) from outside the body, proprioceptors are aware of movements derived from muscular, tendon, and articular sources (kinesthesis or movements of one's own limbs and body parts), and interoceptors transmit information from the internal organs as, for instance, impression from emotions or feelings.

The term *kinesthesis* is often used as a synonym for *proprioception*, e. g. by Robles-De-La-Torre [130, p. 27], but sometimes is referred to as a separate concept [490]. In the latter case, *kinesthesis* is defined as *proprioception* supplemented with *equilibrioception* (or the *vestibular* sense, see "Balance" on p. 215).

In the research context of Sherrington, proprioception (or kinesthesia; both terms are used synonymously) was introduced as an additional sense complementing the classical senses vision, hearing, touch, smell, and taste. Today, this classification is obsolete as it is known that proprioception (in the joints and the inner ear) is not limited to information delivery about one's own activity (active experience), but can also perceive movements forced from outside the body (passive exploration) [489, p. 57]. As a consequence, the theory on the interrelationship of senses has to be abandoned due to the fact that a human's experience ist not clearly attributable to specific receptors or nerves[100].

According to James J. Gibson (1966) [489, pp. 146], kinesthesis refers to movements of the entire body and furthermore strongly resembles *proprioception*[101]. Moreover, the term *kinesthesis* can be subdivided into (*i*) kinesthesia of limbs/joints (movements of the body/torso), (*ii*) vestibulary kinesthesia for indicating movements of the head, (*iii*) cutaneous kinesthesia for the movements of the skin relative to a touched surface, and (*iv*) visual kinesthesia for perspective transformations of the field of view. As set forth by Cholewiak [18], proprioception is respon-

[100] Clinical aspects of proprioception are measured in tests that measure a subject's ability to detect an externally imposed passive movement, or the ability to reposition a joint to a predetermined position [491], [492].

[101] Unfortunately, experimental evidence suggests that there is no strong relationship between these two aspects and furthermore, that they seem to be separated physiologically (however, they may be related in a cognitive manner) [490].

Proprioception (Sense of the position and movement of the body and body-parts)			
SENSORY MODALITY	PERCEPTION	ORGAN	IN-CAR DEVICE OR APPLICATION
Proprioception	Position, Movement of the body	Muscles, Tendons, Joints of bones	Adding active experience to the passive tactile sensitivity
Kinesthesis	Position, Movement of body parts	Limbs and other mobile parts of jointed skeleton	Adding active experience to the passive tactile sensitivity

Fig. C.3: Proprioception and kinesthesis, responsible for reception of stimuli produced within the body.

sible for body position and movement in space, while kinesthesis is the awareness of limb/joint position and movement in space. Body position and movement sensing (proprioception) employs the vestibular sense for increasing accuracy.

With respect to its origin, proprioception (from Latin *proprius*, own, individual, personal and *ception*, awareness, consciousness) is the awareness of one's own body, and kinesthesia (from Greek *kineo*, to move) is the awareness of body part movement and position [398, p. 92]. Following this classification, the term proprioception is situated "above" kinesthesia because proprioception is more perceptual whereas kinesthesia generally refers to raw sensations [18].

According to Schiffman, [398, p. 371], [398, p. 376] and Hayward *et al.* [347, p. 21], *kinesthesis* is the reception of body part position and movement (of the limbs and other mobile parts of the jointed skeleton) in space, and *proprioception* is the class of sensory information arising from vestibular and kinesthetic stimulation. Movement and position stimulation occurs in the joints of bones by activating the Pacinian corpuscles (PC) as responsible cutaneous mechanoreceptors. In contrast, the vestibular senses involve the semicircular and otolith organs, sensing head or whole body acceleration of one type or another – linear or rotatory movement [18].

Freyberger *et al.* [493], Voisin *et al.* [494] or Cholewiak [18] interpret *somesthesis* as sense including *cutaneous* sensations (tactile or contact information) as well as the capability to sense *position* and *movement* of body limbs (*proprioception*). For instance, for the hand, *tactile* feedback could be provided by vibration motors (as used for experiments in the scope of this research work) and *proprioceptive* feedback by using a resistive force feedback exoskeleton[102] to allow users to feel the size and shape of virtual objects [153].

Summary

In this work, somesthesis is indicated as the superordinate term involving both passive experience and active exploration – the first class involves only the sense of touch, the second involves

[102] Immersion Corp. offers a haptic interface for the hand called "CyberGrasp", http://www.inition.co.uk/inition/product.php?URL_=product_glove_vti_grasp&SubCatID_=26 or http://www.immersion.com/products/index.html, last retrieved August 31, 2009.

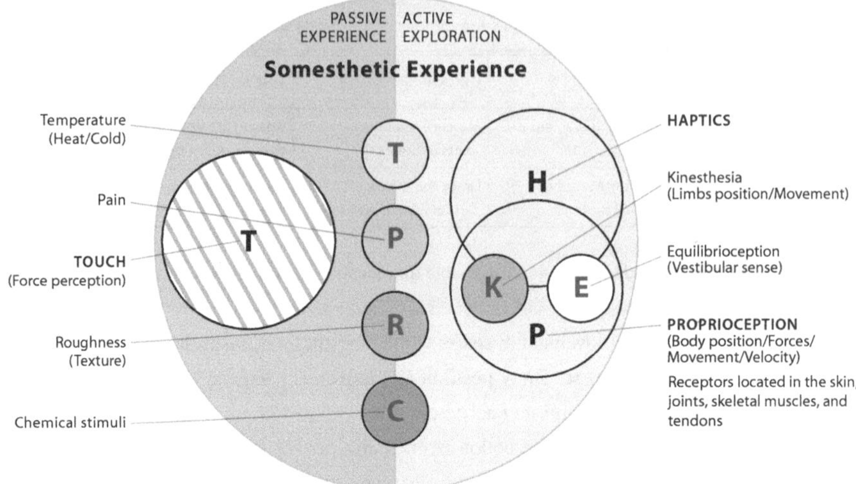

Fig. C.4: Proprioception refers to both kinesthetic and vestibular sensitivity (inspired by Cholewiak [18], [19]).

haptics and proprioception (see Fig. C.4). The four stimuli *temperature*, *pain*, *roughness*, and *chemical stimuli* overlap both sides of the circle [398, p. 367] – they can be used in either passive touch as well as in active exploration [19].

The experiments on the perception of tactile stimuli, the processing of vibro-tactile stimulation, and the utilization of tactors – heavily stressed in this research work – all originate from the sense of touch (situated in the left semicircle in Fig. C.4) and are passive sensations from a stimulation of the skin, such as pressure or touch.

The overlapping circles representing temperature, pain, roughness and chemical stimuli and the right semicircle incorporating active exploration (haptic perception and/or proprioception) (as indicated in Fig. C.4) have been fully disregarded in this work.

D Human Factors

Each interactive system is only as good as its weakest part. In a Driver-2-Vehicle (D2V) interface, the maximum time consumed by the components responsible for sensing, processing, and feedback is determined by the system specifications (and almost unchangeable after integration and installation into the vehicle). Thus irregularities in the system performance (e. g. the time for one cycle in the feedback loop) primarily depend on the sum of the times required by a driver for detecting state changes and reacting accordingly.

In this section, a short summary of human factors influencing the reaction performance is given. The objective is to improve the performance of the response (or feedback) loop in Driver-2-Vehicle interfaces. It has been shown by different researchers that females need on average more time for a response than males (this difference has already been detected in early infancy, e. g. by Surnina and Lebedeva [495]). However, it has been indicated that the male persons make more mistakes than the female individuals (the latter operate more accurately) [441]. Another factor influencing the response time is age – according to several studies, e. g. by Brammer *et al.* [432] or Surnina *et al.* [495], the response time rises almost linear with age.

The following subsections consider all of these parameters (reaction/response time, gender dependency, and age deterioration) separately, give recommendations for improving the performance of affected systems, and furthermore indicate restrictions to be considered when conducting user studies in order to obtain meaningful results.

D.1 Human Reaction Times

Investigating human response times has been of scientific interest for 70 years, but only in the last decades has increasing effort been attempted and reported for this topic, particularly in the automotive field (the term "response time" is here referred to as "driver reaction time"). Much of the research regarding reaction times stems from the early work of Robert Miller (and his 1968 paper on performance analyzation [356]). For example, he proposed an ideal response time of about 2 seconds which is essential to know when designing user interfaces (this value was confirmed in 1974 by Testa *et al.* [355, p. 63]). In the meantime, it has been substantiated that average simple reaction times for light and sound stimuli are below 200*msec* and that sound stimuli are perceived approximately 20% faster than light stimuli, e. g. by Kosinski [496], Galton [497], Fieandt *et al.* [498], Welford [499].

The definition of Human Reaction Time (HRT) (or response time) in this work follows that given by Alan Dix [500] as "Human Reaction Time is defined as the time it takes a human to react to an event such as a light or a sound [..]", and is extended to not be limited to visual or auditory stimulation channels. Shneiderman *et al.* [20] and Teal and Rudnicky [21] have defined the term reaction time from the computer's view as "The computer system's response time is the number of seconds it takes from the moment users initiate an activity until the computer begins to present results [..]", as depicted in Fig. D.1. Furthermore, they assumed that a system delay has an effect on user performance and that this effect can be proven through increased user productivity at decreased system response times ([501, p. 4], [20]).

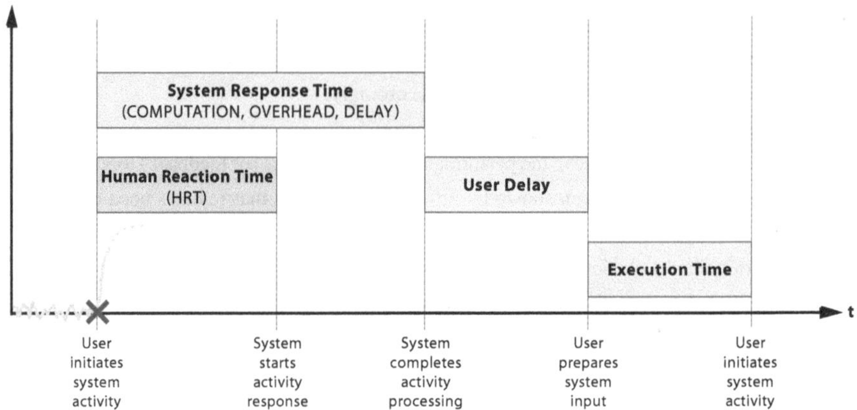

Fig. D.1: Model of system response time (adapted from Shneiderman *et al.* [20], Teal and Rudnicky [21]).

D.1.1 Alternatives and Reaction Time

In [502, p. 4], Triggs and Harris mentioned that the Human Reaction Time depends (almost linearily) on the number of possible alternatives that can occur. As a result, cognitive activity is also limited to a small number of items at any one time. Testa *et al.* [355] discovered that this number is between five and nine and that a human being responds by grouping, sequencing, or neglecting items if more than 5–9 are present. The limited capacity of information absorption for people had already been identified earlier, e. g. in George Miller's paper "The magical number seven, plus or minus two" [390]. He found that people can rapidly recognize approximately seven pieces of information at one time and hold them in short-term memory.

Summary

Measuring and interpretation of human reaction times has a long history, however, was particularly investigated in traditional Human-Computer Interfaces with, for example, a keyboard or mouse as input device and a monitor or speaker as output device. Due to the fact that traditional interaction paradigms are very often not suitable for vehicle handling, evaluation of reaction times has to be repeated in cars. In order to prevent accidents and secure test participants, in this work a trace-driven simulation approach was preferred over a classical on-the-road experiment.

Incorporating the findings of temporal behavior for different sensory channels as reported in the Section "Experiments" should help to (*i*) improve the design of new user interfaces for the automotive domain, and (*ii*) specify the optimal way feedback should be delivered to the driving person. Additionally, when notifying multimodally (this means via two or more channels simultaneously) it has to be taken into account that each modality is best suited of a certain notification demand on the one hand, and that a specific modality has its own constraints with regard to response time on the other – this tradeoff between the choice for a specific modality and its behavior regarding reaction time also needs to be considered.

One example of an interface with improvement potential is the vehicle's dashboard where tens to hundreds of control elements, displays, bulbs, etc. can be activated simultaneously or are even active at the same time. With respect to sensory channels and response times, the dashboard should be extended in such a way as to get control of the individual items and modify their status with respect to its importance (e. g. change the color of the feedback from green to red or notify the driver with an additional auditory signal). In this way, the driver's required cognitive load can be kept at low level.

D.2 Age Dependency

There is strong evidence that perception is affected by age, for instance, Surnina *et al.* [495] reported that the reactivity (ability to respond) decreases with age. Sheppard *et al.* [503] found in an extensive literature review that mental speed is slower among elderly adults and young children, and that the relationship between speed and age over time is curvilinear. An especially large effect can be detected when comparing children to adults or younger to older persons.

In particular, age has been demonstrated to adversely affect tactile sensitivity – as compared to younger individuals, older subjects have lower absolute sensitivity to spatial acuity, pressure sensitivity and vibration [195, pp. 24], [504]. Brammer *et al.* [432] studied the rate of threshold change with age and found that threshold mediated by the *Pacinian* mechanoreceptors increases $2.6dB$ per ten years (measurements took place on the fingertips).

In [433], Shaffer and Harrison confirmed that (*i*) human Pacinian Corpuscles (PC) decrease in number with advanced age, (*ii*) vibro-tactile sensitivity involving PC pathways becomes impaired with age, and (*iii*) older adults ($\overline{x}$=68.6 years) required significantly greater amplitudes of vibration to achieve the same sensation-perceived magnitude as younger subjects ($\overline{x}$=23.5 years). Likewise, Smither *et al.* [434] found that older adults experience a decline in the sensitivity of skin and also have more difficulty in discriminating shapes and textures by touch. As response time gains with increasing age when processing ordinary stimuli, this effect should also be discoverable in vehicles – for example, the time required for hitting the brake pedal of a car when the traffic light turns to red should also increase. The last statement was confirmed by analyzing traffic accidents, e. g. by Kallus *et al.* [435] in Finland. The results show a dramatic age-proportional increase of casualties caused by a (i) decreased speed of information processing and (ii) increased reaction time.

However, on the other hand, experience with a specific task (e. g. dexterity of a piano player, characters per minute typed by a secretary, etc.) apparently compensates for the decline with age [436] – this effect should also be observable for common tasks in the automotive domain, e. g. for hitting the brake pedal or changing the gears.

Summary

The correlation between age and response time has been substantiated and thus should be considered, for instance, when determining prerequisites for experiments or user studies. In order to avoid distortion of results caused by age deterioration, the largest group of test persons should be selected out of a narrow range of age. A notable parameter influencing future interaction with vehicles is the rising expectation of life, leading to a general increase in response times.

D.3 Gender Difference

It has been shown that the mean reaction time for male persons is lower than that for females [495]. Sheppard and Vernon [503] analyzed 172 studies regarding the speed of information processing in relationship to human factors. Their results regarding gender differences in mental speed for both adults and children indicated that it appears that males and females have diverse advantages with regard to differently speeded tasks. In particular, males tend to perform better on reaction and/or inspection tasks, while females perform faster on perceptual speed tasks [503, pp. 540]. Kosinkski [496] reported in his elaborate overview on human factors influencing the reaction time that in almost every age group and across the different interaction channels males respond faster than females. These findings have been substantiated, for instance, by Adam *et al.* [437], Dane *et al.* [438], and Der and Deary [439].

Gender-separated response time experiments for different modalities indicated a 15.38% slower response from visual stimulation in favor of males, and 5% slower female reaction times with auditory stimulation [442]. Another study by Engel [443] found a 6.19% lower performance concerning audio feedback for female test participants.

Earlier task experiments, presented e. g. by Speck *et al.* [440] or Surnina *et al.* [495], indicated higher accuracy and (slightly) slower reaction times for female subjects, too. Barral *et al.* [441] confirmed these results – they found that while men were faster than women at aiming at a target, the women were more accurate.

Population Statistics (separated by age, gender and region)					
REGION/YEAR	AGE RANGE	POPULATION BOTH SEXES	POPULATION MALE	POPULATION FEMALE	SEX RATIO
ENTIRE WORLD 2008	Total (all ages)	6,706,992,932	3,376,791,855	3,330,201,077	101.4
	0 - 19	2,433,408,414	1,253,808,542	1,179,599,872	106.3
	20 - 64	3,766,128,849	1,900,174,941	1,865,953,908	101.8
	65+	507,455,669	222,808,372	284,647,297	78.3
LESS DEVELOPED COUNTRIES 2008	Total (all ages)	5,487,802,029	2,784,539,494	2,703,262,535	103.0
	0 - 19	2,155,194,579	1,111,145,232	1,044,049,347	106.4
	20 - 64	3,018,620,849	1,528,678,878	1,489,941,971	102.6
	65+	313,986,601	144,715,384	169,271,217	85.5
MORE DEVELOPED COUNTRIES 2008	Total (all ages)	1,219,190,903	592,252,361	626,938,542	94.5
	0 - 19	278,213,835	142,663,310	135,550,525	105.2
	20 - 64	747,508,000	371,496,063	376,011,937	98.8
	65+	193,469,068	78,092,988	115,376,080	67.7

Fig. D.2: Midyear population by age and sex (Source: U.S. Census Bureau [22]).

Significance

According to the U.S. Census Bureau [22] the world population is nearly evenly distributed between men and women (sex rate of 101.4). With a sex rate of 98.8 (see Table D.2), illustrating that the population of women is a little higher than that of men, this particularly holds true for the 20 to 64 year age group in the more developed (western) countries (this group represents the major part of premium car drivers globally – presuming that Advanced Driver Assistance Systems are mostly integrated into the class of premium cars).

Summary

With respect to the results of the various studies, confirming nearly equal sized groups of females and males, and substantiated differences in the response time according to gender, it is reasonable to design future user interfaces separately for female and male drivers, either by selecting their operation mode automatically according to the detected driver (preferred option), or manufacturing individual vehicles for both females and males.

E Alphabets Related to Touch

This section provides a short overview of established and well-researched alphabets, potentially influencing the definition of a vibro-tactile alphabet.

E.1 Visual and Auditory Alphabets

E.1.1 Morse Code

Samuel Morse and Alfred Vail, 1836.
English letters are assigned to sequences of "dots", "dashes", and "space". The Morse code was first used to encode characters or words for transmitting telegraphic information. Apart from electric pulses, the short and long elements can also be formed by sound pulses or flashing light and the Morse code could be easily represented by vibro-tactile patterns, responsible for notification of a limited amount of information.

E.1.2 Fingerspelling (ASL)

In Fingerspelling, every letter or word of a writing or numeral system is represented using only the hands. These manual alphabets have spread into a number of single or two-handed sign languages. American Sign Language (ASL), one of the most common sign languages, is often used synonymously with the term sign language, but is actually a specification defining more lexical items than the basic sign language [505].

Apart from a visual representation (tracing the shape of letters in the air), Fingerspelling can also be articulated tactually, tracing letters onto the palm surface of a hand.

E.2 Haptic Alphabets

E.2.1 Braille

Louis Braille, 1824.
The Braille alphabet defines a grid of three (in height) by two (in width) points to present letters, numbers or words. With six points, a maximum of $2^6 = 64$ combinations (including space) is possible. The grid size is about 6*mm* by 4*mm* and was selected according to acuity of touch (the minimum threshold distance at the fingers is about 2*mm*, as depicted in Fig. 10.2 on p. 93). By moving the finger over the raised surface a reading rate of up to 50 words per minute WPM can be achieved [398, p. 102].

E.2.2 Tadoma

Sophie Alcorn, ≈1850.

Tadoma is a method of communication for deaf and blind people and often is referred to as "tactile lip reading". A person places their thumb on the speaker's lips and their fingers along the jawline, touching the speaker's lips and throat.

E.2.3 Vibratese Language

Frank A. Geldard, 1957.

The Vibratese language is a vibratory communication system developed by Geldard [23] in the 1950s at the University of Virginia. Technically, the system consists of five vibration elements mounted on the chest as shown in Fig. E.1, and activated in three levels of intensity ("soft", "medium", and "hard", within $20\mu m$ to $400\mu m$ of stroke) and three duration times (0.1, 0.3, and $0.5 sec$) [398, p. 103]. The vibration frequency was fixed at $60Hz$ for all 45 different elements. The Vibratese language was designed to transmit single letters and digits as well as the most common English words to a user. With this language, more than 60 words per minute WPM can be entered – this is about three times faster than the interpretation of a Morse code (the military standard is 18 words per minute, [398, p. 103]) and a little better than the reading rate of the Braille language.

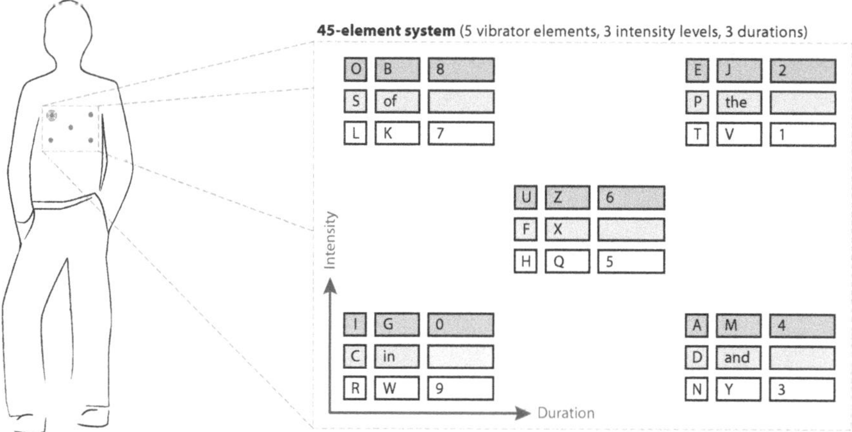

Fig. E.1: Coding of the five vibration elements to 45 letters, digits, and words in the Vibratese language (adapted from Geldard [23]).

E.2.4 Optohapt, Optophone

Frank A. Geldard, 1966.

As the reception of Vibratese was limited by the speed of the transmitting equipment, Geldard experimented with systems for increasing the potential input of (printed) characters and introduced the "Optohapt" system. Basically, the system converts printed text to electrical impulses and in the end to touch stimulations on the body by using nine vibration elements. It represents the symbols directly on the skin surface[103]; thus it is as if the Optohapt "writes" on the skin surface with the vibrators [398, p. 103–104].

Aside from Optohapt (OPtical-TO-HAPTics), other *pictorial, tactual communication systems* have been developed, for instance (*i*) the Optacon (OPtical-to-TActile CONverter), (*ii*) the TVSS (Tactile Vision Substitution System), and (*iii*) the Kinotact (KINesthetic, Optical and TACTile display) [192, p. 582].

[103] For example, writing the letter "V" produces a rapid sweep from the topmost vibrator to the bottom one, and then reverses the sequence.

Bibliography

[1] ETSI. Human Factors (HF); Guidelines on the multimodality of icons, symbols and pictograms. ETSI EG 202 048 DEG/HF-00027, European Telecommunications Standards Institute, ETSI, 650 Route des Lucioles, F-06921 Sophia Antipolis Cedex, France, August 2002.

[2] Thomas A. Stoffregen and Benoit G. Bardy. On specification and the senses. *Behavioral and Brain Science*, 24:195–261, 2001.

[3] Jan B. F. van Erp. *Tactile displays for navigation and orientation: perception and behaviour*. PhD thesis, Utrecht University, The Netherlands and TNO Human Factors, The Netherlands, June 2007. ISBN: 978-90-393-4531-3.

[4] Hong Z. Tan. Perceptual User Interfaces: Haptic Interfaces. *Communications of the ACM*, 43(3):40–41, March 2000.

[5] James F. Allen. Maintaining Knowledge About Temporal Intervals. *Communications of the ACM*, 26(11):832–843, November 1983.

[6] Barry H. Kantowitz and Joseph M. Moyer. Integration of Driver In-Vehicle ITS Information. Technical report, Battelle Human Factors Transportation Center, 4000 NE 41st Street, Seattle, Washington 98105, 2000. Prepared for Federal Highway Administration, Turner Fairbank Highway Research Center.

[7] J. L. Campbell, C. Carney, and B. H. Kantowitz. Human Factors Design Guidelines for ATIS/CVO. Technical Report FHWA-RD-98-057, Battelle Human Factors Transportation Center, 4000 NE 41st Street, Seattle, WA 98105, March 1998. Prepared for Office of Safety and Traffic Operations R&D, Federal Highway Administration, 6300 Georgetown Pike, McLean, VA 22101-2296, 261 pages.

[8] Neil Lerner, Denise Benel, and Debra Dekker. Understanding Driver Performance, Variability and Perception of Risk: Driver Hazard Perception Research Plan. Technical Report FHWA-RD-96-014, US Department of Transportation, Federal Highway Administration, Office of Safety and Traffic Operations R&D, 6300 Georgetown Pike, McLean, VA 22101-2296, March 1998. 42 pages.

[9] Richard Bishop. *Intelligent vehicle technology and trends*. Artech House, Inc., 685 Canton Street, Norwood, MA 02062, May 31, 2005. ISBN: 1-58053-911-4.

[10] Rob van der Heijden and Vincent Marchau. Editorial. In *European Journal of Transport and Infrastructure Research, Special Issue: Advanced Driver Assistance Systems: Behavioural implications of some recent developments*, European Journal of Transport and Infrastructure Research. TU Delft, December 2005. 15 pages, ISSN: 1567-7141.

[11] Alois Ferscha *et al.* Context Framework for Mobile Devices (CON). Industrial Cooperation Siemens AG Munich, Institute for Pervasive Computing, JKU Linz, Presentation Documents, March 22, 2007.

[12] Eric R. Kandel, James H. Schwartz, and Thomas M. Jessell. *Principles of Neural Science*. McGraw-Hill Medical, 4th edition, January 2000. ISBN: 978-0-838-57701-1, 1414 pages.

[13] Alberto Gallace, Hong Z. Tan, and Charles Spence. The Body Surface as a Communication System: The State of the Art after 50 Years. *Presence: Teleoperators & Virtual Environments*, 16(6):655–676, December 2007.

[14] Robert H. Gibson. Electrical stimulation of pain and touch. In Dan R. Kenshalo, editor, *First International Symposium on Skin Senses*, pages 223–261, Florida State University, Tallahassee, March 1968. Springfield, Illinois, Charles C. Thomas. WR 102 I61 1966.

[15] Robert H. Gibson. Electrical Stimulation of Pain and Touch Systems. *Nature*, 199:307–308, July 20, 1963. DOI:10.1038/199307b0.

[16] Yung-Ching Liu. Comparative study of the effects of auditory, visual and multimodality displays on drivers' performance in advanced traveller information systems. *Ergonomics*, 44:425–442(18), March 2001.

[17] Stefan Silbernagel. *Taschenatlas der Physiologie*. Thieme, Stuttgart, 1979. ISBN: 3-135-67707-9, 441 pages.

[18] Roger Cholewiak. Re: Haptics, Tactile, Proprioception. E-Mail, received Tue, 09 Dec 2008 11:41:40 -0500, December 9, 2008. rcholewi@Princeton.EDU, Cutaneous Communication Laboratory, Princeton University, New Jersey, USA.

[19] Roger Cholewiak. Re: Haptics, Tactile, Proprioception. E-Mail, received Fri, 19 Dec 2008 09:14:02 -0500, December 19, 2008. rcholewi@Princeton.EDU, Cutaneous Communication Laboratory, Princeton University, New Jersey, USA.

[20] Ben Shneiderman and Catherine Plaisant. *Designing the User Interface: Strategies for Effective Human-Computer Interaction.* Addison-Wesley, Reading, MA, 3rd edition, 1998. ISBN: 0-201-69497-2.

[21] Steven L. Teal and Alexander I. Rudnicky. A performance model of system delay and user strategy selection. In *CHI '92: Proceedings of the SIGCHI conference on Human factors in computing systems*, pages 295–305, New York, NY, USA, 1992. ACM.

[22] U.S. Census Bureau. International Data Base (IDB). 4600 Silver Hill Road, Washington, DC 20233. Online, last retrieved August 29, 2009. http://www.census.gov/ipc/www/idb/tables.html.

[23] Frank A. Geldard. Adventures in tactile literacy. *The American Psychologist*, 12:115–124, 1957. Report date: September 30, 1956.

[24] Hewett, Baecker, Card, Carey, Gasen, Mantei, Perlman, Strong, and Verplank. ACM SIGCHI Curricula for Human-Computer Interaction, CHAPTER 2: Human-Computer Interaction. Online, last retrieved August 29, 2009 (last document update July 27, 2009). http://www.sigchi.org/cdg/cdg2.html.

[25] Gerhard Mauter and Sabine Katzki. The Application of Operational Haptics in Automotive Engineering. Business Briefing: Global Automotive Manufacturing & Technology 2003 pp. 78–80, Team for Operational Haptics, Audi AG, 2003.

[26] Brad A. Myers. A Brief History of Human Computer Interaction Technology. In *ACM interactions*, volume 5, pages 44–54. Human Computer Interaction Institute, School of Computer Science, Carnegie Mellon University, Pittsburgh, PA 15213-3891, March 1998.

[27] Ben Shneiderman and Catherine Plaisant. *Designing the User Interface: Strategies for Effective Human-Computer Interaction.* Pearson Education, Inc., Addison-Wesley Computing, 4th edition, 2005. ISBN: 0-321-19786-0.

[28] Steven Heim. *The Resonant Interface: HCI Foundations for Interaction Design.* Addison-Wesley Longman Publishing Co., Inc., Boston, MA, USA, March 2007. ISBN:0-321-37596-3.

[29] Hermann Kopetz. *Real-Time Systems – Design Principles for Distributed Embedded Applications.* Kluwer Academic Publishers, 1997.

[30] Rainer Steffen, Richard Bogenberger, Joachim Hillebrand, Wolfgang Hintermaier, Andreas Winckler, and Mehrnoush Rahmani. Design and Realization of an IP-based In-car

Network Architecture. In *The First Annual International Symposium on Vehicular Computing Systems (ISVCS 2008)*, ACM Digital Library, July 22-24, 2008, Trinity College Dublin, Ireland, July 2008. ACM Digital Library. ISBN: 978-963-9799-27-1.

[31] Marko Wolf, André Weimerskirch, and Thomas Wollinger. State of the art: Embedding security in vehicles. *EURASIP Journal on Embedded Systems*, 2007:16, 2007. Article ID 74706.

[32] Paul Green. *The Human-Computer Interaction Handbook: Fundamentals, Evolving Technologies and Emerging Applications*, chapter Motor vehicle driver interfaces, pages 844–860. Lawrence Erlbaum Associates, Inc., Mahwah, NJ, USA, 2nd edition, 2003. ISBN: 978-20-8058-5870-9.

[33] Michael Pettitt, Gary Burnett, and Alan Stevens. An extended keystroke level model (KLM) for predicting the visual demand of in-vehicle information systems. In *Proceedings of the SIGCHI conference on human factors in computing systems (CHI '07)*, pages 1515–1524, New York, NY, USA, 2007. ACM. ISBN: 978-1-59593-593-9.

[34] Leonard Evans. *Traffic Safety and the Driver.* Van Nostrand Reinhold, Thomson Publishing, Inc., 1991. ISBN: 0-442-00163-0.

[35] Caterina Caleefato, Roberto Montanari, and Fabio Tango. Advanced Drivers Assistant Systems in Automation. 2007. Pages 768–777(9).

[36] Morten Meilgaard, Gail Vance Civille, and B. Thomas Carr. *Sensory Evaluation Techniques, 3rd Edition.* CRC Press LLC, Boca Raton, Florida, June 24, 1999. ISBN: 0-8493-0276-5, 416 Pages.

[37] Albrecht Schmidt. Implicit human computer interaction through context. *Personal and Ubiquitous Computing*, 4(2):191–199, June 2000.

[38] Wendy Ju, Brian A Lee, and Scott R Klemmer. Range: Exploring implicit interaction through electronic whiteboard design. In *Proceedings of the 2007 Conference on Human Factors in Computing Systems, CHI 2007, San Jose, California, USA*, page 10, Stanford University, 353 Serra Mall, Stanford CA 94305, April 2007. ACM.

[39] Wendy Ju. *The Design of Implicit Interactions (draft).* PhD thesis, Stanford University, Stanford, CA, 2008.

[40] D. Scott McCrickard, C. M. Chewar, Jacob P. Somervell, and Ali Ndiwalana. A model for notification systems evaluation – assessing user goals for multitasking activity. *ACM Transactions on Computer-Human Interaction (TOCHI'03)*, 10(4):312–338, 2003.

[41] Silvia Schiaffino and Analía Amandi. User - interface agent interaction: personalization issues. *International Journal of Human-Computer Studies*, 60(1):129–148, January 2004.

[42] Niels Ole Bernsen. Defining a taxonomy of output modalities from an hci perspective. *Computer Standards & Interfaces*, 18(6-7):537–553, December 1997.

[43] Joseph Luk, Jerome Pasquero, Shannon Little, Karon MacLean, Vincent Levesque, and Vincent Hayward. A role for haptics in mobile interaction: initial design using a handheld tactile display prototype. In *CHI '06: Proceedings of the SIGCHI conference on Human Factors in computing systems*, pages 171–180, New York, NY, USA, 2006. ACM.

[44] A. Wilson and N. Oliver. Multimodal Sensing for Explicit and Implicit Interaction. In *Proceedings of the 11th International Conference on Human-Computer Interaction (HCII'05)*, Mahwah, NJ, 2005. Lawrence Erlbaum Associates, Las Vegas, Nevada.

[45] Jörn Hurtienne and Lucienne Blessing. Metaphors as Tools for Intuitive Interaction with Technology. *Metaphor in Science and Technology*, page 32, December 2007. ISSN 1618-2006 (Internet), ISSN 1865-0716 (Print).

[46] Carsten Mohs, Jörn Hurtienne, Johann Habakuk Israel, Anja Naumann, Martin Christof Kindsmüller, Herbert Meyer, and Anna Pohlmeyer. IUUI - Intuitive Use of User Interfaces. In *Workshop on User Experience I, Mensch und Computer 2006, Gelsenkirchen, Germany*, September 3–6, 2006. 10 pages.

[47] Anja Naumann, Jörn Hurtienne, Johann Israel, Carsten Mohs, Martin Kindsmüller, Herbert Meyer, and Steffi Husslein. Intuitive Use of User Interfaces: Defining a Vague Concept. In *Engineering Psychology and Cognitive Ergonomics*, volume 4562 of *Lecture Notes in Computer Science (LNCS)*, pages 128–136. Springer Berlin/Heidelberg, August 2007. ISBN: 978-3-540-73330-0.

[48] Anton Nijholt. Multimedia and Interaction. Strategic Research Agenda (SRA) 5, Netherlands Institute for Research on ICT (NIRICT), March 2007.

[49] B. L. Hills. Vision, mobility, and perception in driving. *Perception*, 9:183–216, 1980.

[50] Markus Dahm. *Grundlagen der Mensch-Computer-Interaktion*. Pearson Education, 1st edition, December 2005. 368 pages, ISBN: 978-3-8273-7175-1.

[51] M. Wierda and J. Aasmann. Seeing and driving: computation, algorithms and implementation. *Traffic Research Centre, University of Groningen, The Netherlands*, 1992.

[52] M.C. McCallum, J.L. Campbell, J.B. Richman, J.L. Brown, and E. Wiese. Speech recognition and in-vehicle telematics devices: Potential reductions in driver distraction. *International Journal of Speech Technology*, 7(1):25 – 33, January 2004.

[53] Engin Erzin, Yucel Yemez, A. Murat Tekalp, Aytul Ercil, Hakan Erdogan, and Huseyin Abut. Multimodal person recognition for human-vehicle interaction. *IEEE MultiMedia*, 13(2):18–31, 2006.

[54] Margarita Osadchy and Daniel Keren. Image detection under varying illumination and pose. *Proceedings of Eighth IEEE International Conference on Computer Vision (ICCV'01)*, 2:668–673, 2001.

[55] L. Torres. Is there any hope for face recognition? (position paper). In *International Workshop on Image Analysis for Multimedia Interactive Services (WIAMIS 2004), Lisboa, Portugal*, April 21–23, 2004.

[56] Sharon Oviatt. Ten myths of multimodal interaction. *Communications of the ACM*, 42(11):74–81, 1999.

[57] Dorothy Rachovides, Zoe Swiderski, and Alan P. Parkes. Metaphors and Multimodal Interaction. In *1st International Conference on Computational Semiotics for Games and New Media (COSIGN'01), CWI, Amsterdam*. Computing Department, Lancaster University, Lancaster, UK, September 2001.

[58] Roman Vilimek and Alf Zimmer. Development and evaluation of a multimodal touchpad for advanced in-vehicle systems. In *HCI (13)*, pages 842–851, 2007.

[59] Christopher D. Wickens. Processing resources in attention. *Varieties of Attention, R. Parasuraman (ed.)*, pages 63–97, 1984. New York, Academic Press.

[60] Susan McRoy. Detecting, Repairing, and Preventing Human-Machine Miscommunication. *Artificial Intelligence Magazine*, 18(1):1, Spring 1997. Workshop Report.

[61] Susan McRoy. Detecting, Repairing, and Preventing Human-Machine Miscommunication, Thirteenth National Conference on Artificial Intelligence (AAAI-96), Workshop Program. January 1996.

[62] Robert Kass and Tim Finin. Modeling the User in Natural Language Systems. *Computational Linguistics*, 14(3):5–22, January 1988.

[63] W. Piechulla, C. Mayser, H. Gehrke, and W. König. Online-Fahrerbeanspruchungsschätzung. *Proceedings of 38. BDP-Kongress für Verkehrspsychologie*, September 12–14, 2002. 12 pages.

[64] Fabio Tango and Roberto Montanari. Shaping the drivers' interaction: how the new vehicle systems match the technological requirements and the human needs. *Cognition, Technology & Work*, 8(3):215–226, September 2006.

[65] G. Jahn, A. Oehme, J. Krems, and C. Gelau. Peripheral detection as a workload measure in driving: Effects of traffic complexity and route guidance system use in a driving study. In *Transportation Research Part F 8*, pages 255–275. Department of Psychology, Chemnitz University of Technology, D-09107 Chemnitz, Germany, Elsevier, 2005.

[66] Christopher A. Monk, Deborah A. Boehm-Davis, and J. Gregory Trafton. Recovering from interruptions: Implications for driver distraction research. In *Human Factors*, volume 46, pages 650–663. SAIC, Office of Human-Centered Research, National Highway Traffic Safety Administration, 400 7th St. SW, Washington, DC 20590, Human Factors and Ergonomics Society, Winter 2004.

[67] Driver distraction trends and issues. *Computing & Control Engineering Journal*, 16(1):28–30, February–March 2005. ISSN: 0956-3385.

[68] VicRoads Information Services, PO Box 1644, Melbourne, VIC 3001. Facing the real challenges of driving. Online, last retrieved August 29, 2009. http://www.vicroads.vic.gov.au/NR/rdonlyres/2BB48022-ADED-4EA8-8D0A-6E5BB7668E4D/0/RSPart1.pdf.

[69] Gordon E. Moore. Cramming more components onto integrated circuits. *Electronics*, 38(8):4, April 19, 1965.

[70] Paul S. Otellini. Intel Silicon Innovation: Fueling New Solutions for the Digital Planet. Technical report, Intel Corporation, January 2005.

[71] Robert Bosch GmbH. Product Overview: Automotive Electronics. Online, last retrieved August 28, 2009. http://rb-k.bosch.de/en/service/seekingandfindingaproduct/automotiveelectronics.html.

[72] J. P. Thalen. ADAS for the Car of the Future: Interface Concepts for Advanced Driver Assistant Systems in a Sustainable Mobility Concept of 2020. Technical report, Faculty of Engineering Technology / Industrial Design, University of Twente, Netherlends, April/June 2006.

[73] Jim Newkirk. Euro comfort control system can become tech's 'headache'. Online, last retrieved August 29, 2009, published: October 11, 2006. http://www.asashop.org/autoinc/oct2006/techtips.htm.

[74] M. Joseph Moyer. Human Factors Research Needs for the Intelligent Vehicle Initiative (IVI) Program. Summary Report FHWA-RD-98-147, U.S. Department of Transportation, Federal Highway Administration, Battelle Human Factors Transportation Center, Seattle, Washington, September 1998.

[75] Marcus Toennis, Verena Broy, and Gudrun Klinker. A Survey of Challenges Related to the Design of 3D User Interfaces for Car Drivers. In *Proceedings of the 3D User Interfaces (3DUI'06)*, pages 127–134, Washington, DC, USA, 2006. IEEE Computer Society.

[76] Juergen Schwarz et al. Code of Practice for the Design and Evaluation of ADAS. PReVENT Report 11300, v1.6, Response 3, a PReVENT Project, August 2006. Preventive and Active Safety Applications, Integrated Project, Contract number FP6-507075.

[77] T. A. Dingus, S. G. Klauer, V. L. Neale, and Petersen A. et al. The 100-Car Naturalistic Driving Study: Phase II - Results of the 100-Car Field Experiment. Interim Report DOT HS 810 593, US Department of Transportation, National Highway Traffic Safety Administration (NHTSA), Virginia Tech Transportation Institute, 3500 Transportation Research Plaza (0536), Blacksburg, Virginia 24061, April 2006. 856 pages.

[78] Iowa Department of Transportation. You're the coach! A Guide for Parents of New Drivers, Office of Driver Services, P.O. Box 9204, Des Moines, IA 50313, June 2008.

[79] Jane Stutts, John Feaganes, Eric Rodgman, Charles Hamlett, Thomas Meadows, Donald Reinfurt, Kenneth Gish, Michael Mercadante, and Loren Staplin. Distractions in Everyday Driving. Technical report, University of North Carolina at Chapel Hill and Highway Safety Research Center and TransAnalytics, LLC, June 2003. Funded by AAA Foundation for Traffic Safety, 607 14th Street, NW, Suite 201, Washington, DC 20005.

[80] Andrew Knight. Statistical Bulletin Transport Series Trn/2006/3: Scottish Household Survey Travel Diary Results for 2004. Technical report, The Scottish Government, Transport Statistics, Scottish Executive, Victoria Quay, Edinburgh EH6 6QQ, March 2006. ISSN: 1351-3869, ISBN: 0-7559-2992-6.

[81] Marika Engström. The Swedish Transport Sector Today - Patterns of Travel and Transport. SIKA Rapport 1998:3, Swedish Institute for Transport and Communications, June 1998.

[82] E. Ellis, A. Castle, J. Morewook, K. Farrer, R. Boncinelli, and J. Wharf. Transport Statistics Greater Manchester 2005. GMTU Report 1138, Association of Greater Manchester

Authorities, Greater Manchester Transportation Unit, Salisbury House, Granby Row, Manchester, M1 7AH, June 2005.

[83] Jed Kolko. What to Expect from California's New Hands-Free Law. Occasional papers, Public Policy Institute of California, 500 Washington Street, Suite 600, San Francisco, California 94111, May 2008.

[84] Donna Glassbrenner and Tony Jianqiang Ye. Driver cell phone use in 2006 – overall results. Traffic Safety Facts, Research Note DOT HS 810 790, NHTSA's National Center for Statistics and Analysis, 1200 New Jersey Avenue SE., Washington, DC 20590, July 2007.

[85] Insurance Information Institute. Encyclopedia II - List of Human Senses. Online, last retrieved August 29, 2009. http://www.experiencefestival.com/a/Sense_-_List_of_Human_senses/id/599255.

[86] Terry C. Lansdown, Nicola Brook-Carter, and Tanita Kersloot. Distraction from multiple in-vehicle secondary tasks: vehicle performance and mental workload implications. In *Ergonomics*, volume 47, pages 91–104. Heriot Watt University, Edinburgh, Scotland and Transport Research Laboratory, Crowthorne, UK, Taylor & Francis Ltd., January 2004. ISSN 0014-0139 print/ISSN 1366-5847 online.

[87] Praveen Chandrasekar. Wireless Communication Technologies – Hi-Fi with Wi-Fi and Bluetooth. Online, last retrieved August 29, 2009 (published online October 24, 2005). Frost & Sullivan Market Insight, http://www.frost.com/prod/servlet/market-insight-print.pag?docid=51683460.

[88] Johannes Dünnwald. Nomadic devices – Benefits and market outlook. In *First European Nomadic Devices Workshop*. Nokia Automotive, January 20, 2005.

[89] S. Kiar. Public Perceptions of Road Safety in Canada. Technical report, Transport Canada, Road Safety and Motor Vehicle Regulation, Ottawa, Ontario, 1998.

[90] Ward Vanlaar, Herb Simpson, Dan Mayhew, and Robyn Robertson. The Road Safety Monitor (RSM) 2006: Driver Distraction. Technical report, Traffic Injury Research Foundation (TIRF), 171 Nepean Street, Suite 200, Ottawa, Ontario K2P 0B4, August 2007. ISBN: 978-0-920071-67-0.

[91] Jo Ann Kelley (Coordination). Countermeasures That Work: A Highway Safety Countermeasure Guide For State Highway Safety Offices. Guidelines DOT HS 809 980, U.S. Department of Transportation, National Highway Traffic Safety Administration, 400 Seventh Street, S.W., Washington, DC 20590, January 2006. 190 pages.

[92] D. Glassbrenner. Driver Cell Phone Use in 2005 – Overall Results, Traffic safety facts. Research note DOT HS 809 967, NHTSA's National Center for Statistics and Analysis, NHTSA, 400 Seventh Street SW Washington, DC 20590, 2005.

[93] M. Sundeen. Cell Phone and Highway Safety: 2001 State Legislature Update. In *National Conference of State Legislatures, USA, 2001*, 2001.

[94] H. Alm and L. Nilsson. Changes in driver behaviour as a function of handsfree mobile phones: a simulator study. *Accident Analysis and Prevention*, 26:441–451, 1993.

[95] An investigation of the safety implications of wireless communication in vehicles. Technical report, National Highway Traffic Safety Administration, Department of Transport (NHTSA), NHTSA, Department of Transport, Washington, DC, 1997. Online, last retrieved August 24, 2009. http://www.nhtsa.dot.gov/people/injury/research/wireless/.

[96] D. A. Redelmeier and R.J. Tibshirani. Association between cellular-telephone calls and motor vehicle collisions. *New England Journal of Medicine*, 336:453–458, 1997.

[97] Zosia Bielski. Ontario bill would ban cellphone use by drivers. The Globe and Mail, CTVglobemedia Publishing Inc., 444 Front St. W., Toronto, ON Canada M5V 2S9, October 27, 2008.

[98] L. Morrison-Allsopp. The risk of using a mobile phone while driving. Technical report, Royal Society for the Prevention of Accidents (RoSPA), Birmingham B5 7ST, Edgbaston Park 353 Bristol Road, 2002.

[99] D.L. Strayer, F.A. Drews, and D.J. Crouch. A comparison of the cell phone driver and the drunk driver. In *Human Factors*, number 2, pages 381 – 391. University of Utah, Salt Lake City, Utah, Summer 2006.

[100] A.T. McCartt and L. Hellinga. Review of Research on the Effects of Drivers' Use of Wireless Phones. Technical report, Insurance Institute for Highway Safety, Arlington, 2005.

[101] Transport Canada. Observational Survey of Cell Phone Use by Drivers of Light Duty Vehicles 2006 - 2007. Fact Sheet TP 2436E RS-2008-02, Road Safety and Motor Vehicle Regulation Directorate, Tower C, Place de Ville, 330 Sparks Street, Ottawa, Ontario K1A 0N5, March 2008.

[102] Fotis Karamitsos. IST Information Day Kay Kai, Overview of Transport & Tourism. European Commission, Information Society DG, Online, last retrieved August 25, 2009. ftp://ftp.cordis.europa.eu/pub/ist/docs/ka1/010703_infoday_call7_b5.ppt.

[103] Research and Markets. ADAS (Advanced Driver Assistance Systems) Report and System Database Package. Market research resource, Guinness Centre, Taylors Lane, Dublin 8, Ireland, June 2008. 48 pages, http://www.researchandmarkets.com/reports/607441.

[104] M. Bernardine Dias. Undergraduate Course: Introduction to Robotics and Artificial Intelligence, at Ashesi University in Accra, Ghana, 2006. Carnegie Mellon University, Pittsburgh, PA. Online, last retrieved August 25, 2009. http://www.ashesi.org/ACADEMICS/compsci/lectures/sensing/sensing_perception.pdf.

[105] Robert F. Schmidt, Florian Lang, and Gerhard Thews, editors. *Physiologie des Menschen mit Pathophysiologie*, volume 29 of *Springer-Lehrbuch*. Springer Medizin Verlag Heidelberg, 2005. ISBN: 978-3-540-21882-3, 994 pages.

[106] Ken Hinckley, R. Jacob, and C. Ware. Input/output devices and interaction techniques. In *CRC Computer Science and Engineering Handbook*. CRC Press, Boca Raton, Florida, 2004.

[107] Patrick Garrigan and Philip J. Kellman. Perceptual learning depends on perceptual constancy. *Proceedings of the National Academy of Sciences*, 105(6):2248–2253, February 2008.

[108] Ching Kung. A possible unifying principle for mechanosensation. *Nature*, 436(7051):647–654, August 2005.

[109] Seppo Pohja. Survey of Studies on Tactile Senses. ISSN : 0283-3638 R960:02, RWCP Neuro Swedish Institute of Computer Science (SICS) Laboratory, Box 1263, S-164 28 Kista, Sweden, March 1996. ISSN: 0283-3638.

[110] Richard L. Doty (Editor). *Handbook of Olfaction and Gustation*. Marcel Dekker, Inc. New York, 2003. Second, Revised and Expanded, Edition, ISBN: 0-8247-0719-2, 1121 Pages.

[111] E. Richard, A. Tijou, P. Richard, and J.-L. Ferrier. Multi-modal virtual environments for education with haptic and olfactory feedback. *Virtual Reality*, 10(3):207–225, December 2006.

[112] Joseph Nathaniel Kaye. *Symbolic Olfactory Display*. Master thesis, Massachusetts Institute of Technology, Media Laboratory, Media Arts and Sciences, May 2001.

[113] Mélanie Aeschlimann. The perception of pain and temperature. *EuroBrain (The European Dana Alliance for the Brain, Centre de Neurosciences Psychiatriques, Dép. Universitaire de Psychiatrie - CHUV, Site de Cery, CH-1008 Prilly / Lausanne)*, 5(1):6, June 2004.

[114] G. F. Gebhart. Scientific Issues of Pain and Distress. In *Proceedings of the Workshop on "Definition of Pain and Distress and Reporting Requirements for Laboratory Animals"*. Department of Pharmacology, University of Iowa, Iowa City, Iowa, National Academy Press, Washington, D.C., June 22, 2000.

[115] Russell K. Portenoy and Michael J. Brennan. *Handbook of Neurorehabilitation*, chapter Chronic Pain Management. Informa Health Care, 1994. ISBN: 0-824-78822-2, 704 pages.

[116] Rainer Schönhammer. Bodies, things and movement - as the sense of balance likes it. In *Colloquium on Psychological Aspects of Material Culture and Design*, Burg Giebichenstein, University of Art and Design, Halle (Saale), Germany, October 25–28, 2007. Society for Cultural Psychology.

[117] Alejandro Jaimes and Nicu Sebe. Multimodal human computer interaction: A survey. In *Computer Vision in Human-Computer Interaction (ICCV-HCI)*, Lecture Notes in Computer Science (LNCS). Springer Verlag, 2005. 15 pages.

[118] Azizuddin Khan, Narendra K Sharma, and Shikha Dixit. Effect of Cognitive Load and Paradigm on Time Perception. In *Journal of the Indian Academy of Applied Psychology*, volume 32, pages 32–47. Indian Institute of Technology, Kanpur, January 2006.

[119] Andrei Voustianiouk and Horacio Kaufmann. Magnetic fields and the central nervous system. *Clinical Neurophysiology*, 111(11):1934–1935, November 2000.

[120] Alexander Hof and Eli Hagen. Help Strategies for Speech Dialogue Systems in Automotive Environments. *Perception and Interactive Technologies*, pages 107–116, 2006. LNAI 4021.

[121] Erica L. Goldman, Erin Panttaja, Andy Wojcikowski, and Robert Braudes. Voice Portals -Where Theory Meets Practice. *International Journal of Speech Technology*, 4(3):227–240, July 2001.

[122] Santosh Basapur, Shuang Xu, Mark Ahlenius, and Young Lee. User Expectations from Dictation on Mobile Devices. *Human-Computer Interaction. Interaction Platforms and Techniques*, pages 217–225, August 2007. ISSN: 0302-9743.

[123] Niels Ole Bernsen. Natural human-human-system interaction, 2001.

[124] Daniel W. Repperger, C. A. Phillips, J. E. Berlin, A. T. Neidhard-Doll, and M. W. Haas. Human-machine haptic interface design using stochastic resonance methods. *IEEE Transactions on Systems, Man, and Cybernetics, Part A*, 35(4):574–582, 2005.

[125] S. Fels, R. Hausch, and A. Tang. Investigation of Haptic Feedback in the Driver Seat. *IEEE Intelligent Transportation Systems Conference (ITSC'06)*, pages 584–589, 2006.

[126] Massimo Cellario. Human-Centered Intelligent Vehicles: Toward Multimodal Interface Integration. *IEEE Intelligent Systems*, 16(4):78–81, 2001.

[127] P. Rani, N. Sarkar, C.A. Smith, and J.A. Adams. Affective communication for implicit human-machine interaction. *IEEE International Conference on Systems, Man and Cybernetics*, 5:4896–4903, Vol. 5, October 2003.

[128] Susan J. Lederman. The sense of touch in humans: Implications for designing haptic interfaces for teleoperation and virtual environments. In *Proceedings of 9th International Conference on Artificial Reality and Teleexistence (ICAT'99)*. Queen's University, Depts. of Psychology and Computing & Information Science, Kingston, Ontario, Canada, December 16–18, 1999. ISSN:1345-1278.

[129] Federico Barbagli, Ken Salisbury, Cristy Ho, Charles Spence, and Hong Z. Tan. Haptic discrimination of force direction and the influence of visual information. *ACM Trans. Appl. Percept.*, 3(2):125–135, 2006.

[130] Gabriel Robles-De-La-Torre. The Importance of the Sense of Touch in Virtual and Real Environments. *IEEE MultiMedia*, 13(3):24–30, 2006.

[131] Ming C. Lin and Miguel A. Otaduy. Sensation-Preserving Haptic Rendering. *IEEE Computer Graphics and Applications*, 25(4):8–11, 2005.

[132] Jan B.F. Van Erp and Hendrik A. H. C. Van Veen. Vibrotactile in-vehicle navigation system. *Transportation Research Part F: Traffic Psychology and Behaviour*, 7(4–5):247–256, 2004.

[133] Cristy Ho, Charles Spence, and Hong Tan. Warning Signals Go Multisensory. In *Proceedings of the 11th International Conference on Human-Computer Interaction (HCII'05)*, volume 9, Mahwah, NJ, 2005. Lawrence Erlbaum Associates, Las Vegas, Nevada.

[134] European Commission, Directorate-General Information Society and Media, ICT for Transport. i2010 Intelligent Car Initiative. BU31 4/66, Avenue de Beaulieu, B-1160 Brussels. Online, last retrieved August 30, 2009. http://ec.europa.eu/information_society/activities/intelligentcar/index_en.htm.

[135] European Commission, Directorate-General Information Society and Media, ICT for Transport. Intelligent Car Brochure. Online, last retrieved August 30, 2009, published:

August 25, 2006. http://ec.europa.eu/information_society/activities/intelligentcar/docs/right_column/intelligent_car_brochure.pdf.

[136] Andrea Heide and Klaus Henning. The "cognitive car": A roadmap for research issues in the automotive sector. In *Annual Reviews in Control*, volume 30, pages 197 – 203. Department of Computer Science in Mechanical Engineering, RWTH Aachen University, Germany, Elsevier, September 2006.

[137] Wendy Ju and Larry Leifer. The design of implicit interactions: Making interactive systems less obnoxious. *Design Issues*, 24(3):72–84, 2008.

[138] Albrecht Schmidt. *Ubiquitous Computing – Computing in Context*. PhD thesis, Computing Department, Lancaster University, UK, Computing Department, Engineering Building, Room A13, Lancaster University, Lancaster, UK LA1 4YR, November 2002.

[139] Robert Graham and Chris Carter. Comparison of speech input and manual control of in-car devices while on the move. *Personal and Ubiquitous Computing*, 4(2/3), 2000.

[140] J. Stallkamp, H. K. Ekenel, H. Erdogan, R. Stiefelhagen, and A. Ercil. Video-Based Driver Identification Using Local Appearance Face Recognition. In *Workshop on DSP in Mobile and Vehicular Systems, Istanbul, Turkey*. Interactive Systems Labs, Department of Computer Science, TU Karlsruhe, Germany, June 2007.

[141] J. C. McCall and M.M. Trivedi. Human Behavior Based Predictive Brake Assistance. *Intelligent Vehicles Symposium, 2006 IEEE*, pages 8–12, June 2006.

[142] Hendrik Witt and Holger Kenn. Towards implicit interaction by using wearable interaction device sensors for more than one task. In *Proceedings of the 3rd international conference on mobile technology, applications & systems (Mobility'06)*, New York, NY, USA, 2006. ACM. 20 pages.

[143] Loureiro Paulo, Adolfo Sachsida, and Tito Moreira. Traffic accidents: an econometric investigation. *Economics Bulletin*, 18(3):1–7, March 2004. http://www.economicsbulletin.com/2004/volume18/EB?04R40001A.pdf.

[144] European Communities. Saving 20,000 lives on our roads – A shared responsibility. Technical report, Luxembourg, Office for Official Publications of the European Communities, 2003. 64 pages, ISBN: 92-894-5893-3, Online, last retrieved August 30, 2009. http://ec.europa.eu/transport/roadsafety_library/rsap/rsap_en.pdf.

[145] Roman Vilimek, Thomas Hempel, and Birgit Otto. Multimodal interfaces for in-vehicle applications. In *HCI (3)*, pages 216–224, 2007.

[146] Andreas Holzinger, Kizito Mukasa, and Alexander Nischelwitzer. Introduction to the special thematic session: Human-computer interaction and usability for elderly (hci4aging). Number 5105 in LNCS, pages 18–21. Springer-Verlag Berlin Heidelberg, 2008. ISBN: 978-3-540-70539-0.

[147] Matthew S. Prewett, Liuquin Yang, Frederick R. B. Stilson, Ashley A. Gray, Michael D. Coovert, Jennifer Burke, Elizabeth Redden, and Linda R. Elliot. The benefits of multimodal information: a meta-analysis comparing visual and visual-tactile feedback. In *ICMI '06: Proceedings of the 8th international conference on multimodal interfaces*, pages 333–338, New York, NY, USA, 2006. ACM.

[148] Jason Pascoe, Nick Ryan, and David Morse. Using while moving: HCI issues in fieldwork environments. *ACM Transactions on Computer Human Interaction*, 7(3):417–437, 2000.

[149] Daniel Siewiorek, Asim Smailagic, and Matthew Hornyak. Multimodal Contextual Car-Driver Interface. *IEEE International Conference on Multimodal Interfaces*, pages 367–373(7), 2002.

[150] Roman Vilimek, Thomas Hempel, and Birgit Otto. Multimodal Interfaces for In-Vehicle Applications. *Human-Computer Interaction. HCI Intelligent Multimodal Interaction Environments*, pages 216–224, 2007.

[151] Roberto Pieraccini, Krishna Dayanidhi, Jonathan Bloom, Jean-Gui Dahan, Michael Phillips, Bryan R. Goodman, and K. Venkatesh Prasad. Multimodal conversational systems for automobiles. *Commun. ACM*, 47(1):47–49, 2004.

[152] Roberto Pieraccini, Krishna Dayanidhi, Jonathan Bloom, Jean-Gui Dahan, Michael Phillips, Bryan R. Goodman, and K. Venkatesh Prasad. A Multimodal Conversational Interface for a Concept Vehicle . In *8th European Conference on Speech Communication and Technology (EuroSpeech'03), Geneva, Switzerland*, pages 2233–2236, September 2003.

[153] Franziska Freyberger, Berthold Farber, Martin Kuschel, and Martin Buss. Perception of Compliant Environments Through a Visual-Haptic Human System Interface. *International Conference on Cyberworlds (CW '07)*, pages 314–321, October 2007.

[154] Georgios Yfantidis and Grigori Evreinov. The Amodal Communication System Through an Extended Directional Input. *Computers Helping People with Special Needs (ICCHP'06)*, LNCS 4061:1079–1086, 2006.

[155] Georgios Yfantidis and Grigori Evreinov. Adaptive blind interaction technique for touch-screens. *Universal Access in the Information Society*, 4(4):328–337, May 2006.

[156] Lawrence W. Barsalou. Perceptual symbol systems. *Behavioral and Brain Sciences, Cambridge University Press*, 22(0140-525X/99):577–660 (33), 1999.

[157] Bence Nanay. Four theories of amodal perception. In *Cognitive Science Journal*, pages 1331 – 1336. Syracuse University, Department of Philosophy, 535 Hall of Languages Syracuse, NY 13244 USA, 2007.

[158] Slobodan Markovich. Amodal completion in visual perception. In *Visual Mathematics*, volume 4. Laboratory of Experimental Psychology, Department of Psychology, University of Belgrade, Filozofski fakultet, Cika Ljubina 18-20, 11000 Belgrade, 2002. 15 p., Online, last retrieved August 29, 2009. http://www.mi.sanu.ac.rs/vismath/fila/index.html.

[159] Monika Schwarz and Jeanette Chur. *Semantik: Ein Arbeitsbuch*. Gunter Narr Verlag, 2004. ISBN 3-823-36085-X.

[160] Robert Lickliter, Lorraine E. Bahrick, and Hunter Honeycutt. Intersensory redundancy enhances memory in bobwhite quail embryos. *Infancy*, 5(3):253–269, 2004.

[161] Robert L. Goldstone and Lawrence W. Barsalou. Reuniting perception and conception. *Cognition, Elsevier Science B.V*, 65(2-3):231–262, January 1998.

[162] Kai Richter and Michael Hellenschmidt. Interacting with the ambience: Multimodal interaction and ambient intelligence. In *W3C Workshop on Multimodal Interaction 2004 (MMI)*, page 6, Sophia Antipolis, France, 19-20 July 2004 2004.

[163] David J. Wheatley and Joshua B. Hurwitz. The Use of a Multi-Modal Interface to Integrate In-Vehicle Information Presentation. In *International Driving Symposium on Human Factors in Driver Assessment, Training and Vehicle Control, Aspen, Colorado*. User Centered Research, Motorola Labs, Schaumburg, Illinois, USA, August 2000. 4 pages.

[164] W. B. Verwey, H. Alm, J. A. Groger, W. H. Janssen, J. J. Juiken, J. M. Schraagen, J. Schumann, W. Van Winsum, and H. Wontorra. *Generic Intelligent Driver Support – A Comprehensive Report on GIDS*, chapter GIDS Functions (chapter 7), pages 113–146. Taylor and Francis, London, UK, September 1993. ISBN: 0-74840-069-9.

[165] Marika Hoedemaeker and Mark Neerincx. Attuning In-Car User Interfaces to the Momentary Cognitive Load. pages 286–293, 2007.

[166] M.C. Hulse, T.A. Dingus, M.A. Mollenhauer, Y. Liu, S.K. Jahns, T. Brown, and B. McKinney. Development of Human Factors Guidelines for Advanced Traveler Information Systems and Commercial Vehicle Operations: Identification of the Strengths and Weaknesses of Alternative Information Display Formats. Technical Report FHWA-RD-96-142, Center for Computer-Aided Design, University of Iowa, October 1998. prepared for U.S. Department of Transportation, Federal Highway Administration Research, Development and Technology, Turner-Fairbank Highway Research Center, 189 pages.

[167] Rob van der Heijden and Vincent Marchau. Advanced Driver Assistance Systems: Behavioural implications of some recent developments (editorial). *European Journal of Transport and Infrastructure Research (EJTIR)*, 4:239–252, December 2005. ISSN: 1567-7141.

[168] Jan B.F. Van Erp and Hendrik A. H. C. Van Veen. Vibro-Tactile Information Presentation in Automobiles. In *Proceedings of EuroHaptics 2001*, pages 99–104, TNO Human Factors, Department of Skilled Behaviour, P.O. Box 23, 3769 ZG Soesterberg, The Netherlands, 2001.

[169] Cornelie van Driel and Bart van Arem. Investigation of user needs for driver assistance: results of an internet questionnaire. *European Journal of Transport and Infrastructure Research (EJTIR), Special issue: Advanced Driver Assistance Systems: Behavioural implications of some recent developments*, 4:397–316, December 2005. Vincent Marchau and Rob van der Heijden (Eds.), ISSN: 1567-7141.

[170] Bill Buxton. A directory of sources for input technologies. Online, last retrieved August 29, 2009 (last update January 29, 2009). http://www.billbuxton.com/InputSources.html, November 2008.

[171] Arjan Geven, Reinhard Sefelin, Manfred Tscheligi, Clemens Kaufmann, and Ralf Risser. Studying UX Aspects of In-Car Vibrotactile Interaction. In *2nd Workshop on Automotive User Interfaces and Interactive Applications (AUIIA'08), Mensch und Computer 2008, Luebeck, Germany*. Logos Verlag, Berlin, September 7–10 2008. ISBN: 978-3-8325-2007-6, 6 pages.

[172] Dagmar Kern, Marco Müller, Stefan Schneegass, Lukasz Wolejko-Wolejszo, and Albrecht Schmidt. CARS – Configurable Automotive Research Simulator. In *2nd Workshop on Automotive User Interfaces and Interactive Applications (AUIIA'08), Mensch und Computer 2008, Luebeck, Germany*. Logos Verlag, Berlin, September 7–10, 2008. ISBN: 978-3-8325-2007-6, 5 pages.

[173] Erik Hollnagel. Cognition as control: A pragmatic approach to the modelling of joint cognitive systems. In *Special Issue of IEEE Transactions on Systems, Man, and Cybernetics*, October 2002. 23 pages, *Note: The special issue was never completed and published.*

[174] P. Carlo Cacciabue, editor. *Modelling Driver Behaviour in Automotive Environments, Critical Issues in Driver Interactions with Intelligent Transport Systems*. Springer, 1st edition, 2007. ISBN: 978-1-84628-617-9, 428 pages.

[175] Martin Pielot, Dirk Ahlers, Wilko Heuten, Amna Asif, and Susanne Boll. Applying Tactile Displays to Automotive User Interfaces. In *2nd Workshop on Automotive User Interfaces and Interactive Applications (AUIIA'08), Mensch und Computer 2008, Luebeck, Germany*. Logos Verlag, Berlin, September 7–10 2008. ISBN: 978-3-8325-2007-6, 3 pages.

[176] Kenneth Majlund Bach, Mads Gregers Jaeger, Mikael B. Skov, and Nils Gram Thomassen. You can touch, but you can't look: interacting with in-vehicle systems. In *Proceedings of the 26th annual SIGCHI conference on human factors in computing systems (CHI'08)*, pages 1139–1148, New York, NY, USA, 2008. ACM.

[177] Sarah Dowdey. How Taste Works. Online, last retrieved August 31, 2009 (published online October 25, 2007). http://health.howstuffworks.com/taste.htm.

[178] Dana M. Small. *Taste and Smell. An Update*, volume 63 of *Advances in Oto-Rhino-Laryngology*, chapter Central Gustatory Processing in Humans, pages 191 – 220. S. Karger AG, Basel, 2006.

[179] Robbie Schäfer, Steffen Bleul, and Wolfgang Müller. *A Novel Dialog Model for the Design of Multimodal User Interfaces*, volume 3425/2005 of *Lecture Notes in Computer Science, Engineering Human Computer Interaction and Interactive Systems*, pages 221–223. Springer Berlin, 2005.

[180] Josef Ulrich. Der erste Opel, der sehen kann! Media Information, General Motors Austria GmbH, Gross-Enzersdorfer Strasse 59, A-1220 Vienna, June 18, 2008.

[181] F. M. Cardullo. Physiological effects of high-G flight - Their impact on flight simulator design. In *Flight Simulation Technologies Conference, Long Beach, CA*, number Technical Papers A81-36554 16-09, pages 147–153. New York, State University, Binghamton, N. Y., American Institute of Aeronautics and Astronautics, Inc., New York, June 16–18, 1981.

[182] Henricus Van Veen and Jan Van Erp. Tactile information presentation in the cockpit. In *Proceedings of First International Workshop on Haptic Human-Computer Interaction, Glasgow, UK*, pages 174–181. Springer Berlin, Heidelberg, August 31 – September 1, 2001. ISBN: 978-3-540-42356-0.

[183] The National Institute for Rehabilitation Engineering (NIRE). Information about "Impaired Night Vision" and "Night Blindness. Technical report, ABLEDATA, 8630 Fenton Street, Suite 930, Silver Spring, USA, 2002. http://www.abledata.com/abledata_docs/Night_Vision.htm.

[184] Mario V. Capanzana. Night Blindness Leaves Ten Out of Every 100 Mothers Groping in the Dark. Food and Nutrition Research Institute (FNRI), Department of Science and Technology (DOST), Gen. Santos Avenue, Bicutan, Taguig City, Metro Manila, Philippines 1631.

[185] P. A. Wilkins and W. I. Action. Noise and accidents – a review. *Annals of Occupational Hygiene*, 25(3):249–260, 1982.

[186] R. L. Brown, W. D. Galloway, and K. R. Gildersleeve. Effects of Intense Noise on Processing of Cutaneous Information of Varying Complexity. *Percept Mot Skills*, 20:749–754, 1965.

[187] Frank A. Geldard. Some Neglected Possibilities of Communication. *Science*, 131(3413):1583–1588, May 27, 1960.

[188] B. Von Haller Gilmer. Possibilities of cutaneous electro-pulse communication. *Symposium on cutaneous sensitivity*, pages 76–84, 1960.

[189] Geoffrey J. Jeram. *Open Platform for Limit Protection with Carefree Maneuver Applications*. PhD thesis, Georgia Institute of Technology, December 2004.

[190] Geoffrey J. Jeram and J. V. Prasad. Open Architecture for Helicopter Tactile Cueing Systems. *American Helicopter Society*, 50(3):238–248, 2005. ISSN: 0002-8711.

[191] Jan B.F. Van Erp, , and Marc H. Verschoor. Cross-modal visual and vibrotactile tracking. *Applied Ergonomics*, 35(2):105–112, March 2004.

[192] Hong Z. Tan and Alex Pentland. *Fundamentals of Wearable Computers and Augmented Reality*, chapter Tactual Displays for Sensory Substitution and Wearable Computers, pages 578–598. Number 105. Lawrence Erlbaum Associates, Mahway, NJ, January 2001. ISBN: 978-0-8058-2902-0.

[193] Hong Z. Tan, Ifung Lu, and Alex Pentland. The chair as a novel haptic user interface. *Proceedings of the Workshop on Perceptual User Interfaces*, pages 19–21, 1997.

[194] Charles G. Gross and Asif A. Ghazanfar. Neuroscience: A mostly sure-footed account of the hand. *Science*, 312(5778):1314, June 2006.

[195] Gary Fryer. *Distinguishing Characteristics of Thoracic Medial Paraspinal Structures Determined as Abnormal by Palpation*. PhD thesis, Victoria University, School Of Health Science, Faculty of Health, Engineering and Science, February 2007.

[196] James C. Craig and Gary B. Rollman. Somesthesis. In *Annual Review of Psychology*, volume 50, pages 305–331, Department of Psychology, Indiana University, Bloomington, Indiana and Department of Psychology, University of Western Ontario, London, Ontario, CD, 1999. Annual Reviews.

[197] Ranjith Kumar Thangavelu. Effect of Non-Visual Stimulus on Color Perception. Master thesis, North Carolina State University, 2003. 154 pages.

[198] Vasilios G. Chouvardas, Amalia N. Miliou, and Miltiadis Hatalis. Tactile Displays: a Short Overview and Recent Developments. In *Proceedings of the 5th International Conference on Technology and Automation (ICTA'05)*, pages 246–251. Department of Informatics, Aristotle University of Thessaloniki, Greece, 2005.

[199] Aaron Toney, Lucy Dunne, Bruce H. Thomas, and Susan P. Ashdown. A Shoulder Pad Insert Vibrotactile Display. *Seventh IEEE International Symposium on Wearable Computers (ISWC'03), White Plains, New York*, pages 35–44, October 21-23, 2003.

[200] Hiroyuki Kajimoto, Naoki Kawakami, Susumu Tachi, and Masahiko Inami. SmartTouch: Electric Skin to Touch the Untouchable. *IEEE Computer Graphics and Applications*, 24(1), 2004. Pages 36–43.

[201] Lynette A. Jones, Brett Lockyer, and Erin Piateski. Tactile display and vibrotactile pattern recognition on the torso. *Advanced Robotics*, 20(12):1359–1374, 2006.

[202] Robert W. Lindeman, Yasuyuki Yanagida, John L. Sibert, and Robert Lavine. Effective vibrotactile cueing in a visual search task. In Matthias Rauterberg, Marino Menozzi, and Janet Wesson, editors, *Proceedings of 13th International Human-Computer Interaction (INTERACT'03)*, pages 89–96. IFIP, IOS Press, 2003. ISBN:1-58603-363-8.

[203] Jan B.F van Erp. Presenting directions with a vibrotactile torso display. *Ergonomics, Taylor and Francis Ltd.*, 48(3):302–313, February 2005.

[204] Hendrik A.H.C Van Veen and Jan B.F. Van Erp. Providing Directional Information with Tactile Torso Displays. In *Eurohaptics 2003*. TNO Human Factors Soesterberg, The Netherlands, Springer LNCS, July 6–9, 2003.

[205] Robert W. Lindeman, Robert Page, Yasuyuki Yanagida, and John L. Sibert. Towards full-body haptic feedback: the design and deployment of a spatialized vibrotactile feedback system. In *Proceedings of the ACM symposium onvVirtual reality software and technology (VRST '04)*, pages 146–149, New York, NY, USA, 2004. ACM.

[206] W. Lindeman, Yasuyuki Yanagida, Haruo Noma, and Kenichi Hosaka. Wearable vibrotactile systems for virtual contact and information display. *Virtual Real.*, 9(2):203–213, 2006.

[207] Linda R. Elliott, Elizabeth S. Redden, Rodger A. Pettitt, Christian B. Carstens, Jan van Erp, and Maaike Duistermaat. Tactile Guidance for Land Navigation. Technical Report ARL-TR-3814, Army Research Laboratory, Aberdeen Proving Ground, MD 21005-5425, 2006.

[208] M. Duistermaat. Tactile Land Navigation in Night Operations. Technical report, TNO Defence, Security and Safety Kampweg 5 3769 ZG Soesterberg, The Netherlands, December 2005. R&D No. 9954-AN-01.

[209] Koji Tsukada and Michiaki Yasumura. Activebelt: Belt-type wearable tactile display for directional navigation. In *Ubicomp*, pages 384–399, 2004.

[210] Jan B. F. van Erp, Hendrik A. H. C. van Veen, Chris Jansen, and Trevor Dobbins. Waypoint navigation with a vibrotactile waist belt. *ACM Trans. Appl. Percept.*, 2(2):106–117, 2005.

[211] Alois Ferscha, Bernadette Emsenhuber, Andreas Riener, Clemens Holzmann, Manfred Hechinger, Dominik Hochreiter, Marquart Franz, Andreas Zeidler, Marcos dos Santos Rocha, and Cornel Klein. Vibro-Tactile Space-Awareness. In *Adjunct Proceedings of the 10th International Conference on Ubiquitous Computing (UBICOMP'08), Seoul, South Korea, Video paper*, September 2008.

[212] Hong Z. Tan and Alex Pentland. Tactual displays for wearable computing. *Personal and Ubiquitous Computing*, 1(4):225–230, December 1997.

[213] Sevgi Ertan, Clare Lee, Abigail Willets, Hong Z. Tan, and Alex Pentland. A wearable haptic navigation guidance system. In *ISWC*, pages 164–165, 1998.

[214] Hong Z. Tan, Robert Gray, J. Jay Young, and Ryan Traylor. A Haptic Back Display for Attentional and Directional Cueing. In *Haptics-e*, volume 3, page 20, Haptic Interface Research Laboratory, Purdue University, 1285 Electrical Engineering Building, West Lafayette, IN, USA 47907, 2003.

[215] Jan B. F. Van Erp, Hendrik A. H. C. Van Veen, Chris Jansen, and Trevor Dobbins. Waypoint navigation with a vibrotactile waist belt. *ACM Trans. Appl. Percept.*, 2(2):106–117, April 2005.

[216] Cristy Ho, Hong Z. Tan, and Charles Spence. Using spatial vibrotactile cues to direct visual attention in driving scenes. In *Transportation Research Part F*, volume 8, pages 397–412. Department of Experimental Psychology, University of Oxford, South Parks Road, Oxford OX1 3UD, United Kingdom and Haptic Interface Research Laboratory, Purdue University, Indiana, USA, Elsevier (Sciencedirect), 2005.

[217] Cristy Ho, Hong Z. Tan, and Charles Spence. The differential effect of vibrotactile and auditory cues on visual spatial attention. *Ergonomics, Taylor and Francis Ltd.*, 49(7):724–738, June 10, 2006.

[218] R. W. Cholewiak, C. E. Sherrick, and A. A. Collins. Studies of saltation. Princeton Cutaneous Research Project 62, Princeton University, Department of Psychology, 1996.

[219] Alberto Gallace, Hong Z. Tan, and Charles Spence. Numerosity judgments for tactile stimuli distributed over the body surface. *Perception 35 Psychophysics*, 35(2):247–266, 2006.

[220] Alberto Gallace, Hong Z. Tan, and Charles Spence. The body surface as a communication system: The state of the art after 50 years. *Presence: Teleoperators and Virtual Environments*, 16(6):655–676, December 2007. ISSN:1054-7460.

[221] Alberto Gallace, Hong Z. Tan, and Charles Spence. Multisensory numerosity judgments for visual and tactile stimuli. *Perception 38; Psychophysics*, 69:487–501, May 2007.

[222] Kirby Gilliland and Robert E. Schlegel. Tactile stimulation of the human head for information display. *Human Factors: The Journal of the Human Factors and Ergonomics Society*, 36:700–717(18), December 1994.

[223] Hiroaki Yano, Tetsuro Ogi, and Michitaka Hirose. Development of haptic suit for whole human body using vibrators. In *Transactions of the Virtual Reality Society of Japan*, volume 3, pages 141–147. Faculty of Engineering, University of Tokyo, 1998.

[224] Wiel Janssen and L. Nilsson. An experimental evaluation of in-vehicle collision avoidance systems. In *Proceedings of the 24 th ISATA Symposium, Florence*, May 1991.

[225] Wiel Janssen and Hugh Thomas. Collision avoidance support under conditions of adverse visibility. Technical Report Report IZF 1993 C-, TNO Institute for Human Factors, Soesterberg, 1993.

[226] H. Godthelp and J. Schumann. *Driving future vehicles*, chapter Intelligent Accelerator: An Element of Driver Support (Chapter 25). Taylor & Francis, London, 1993.

[227] W. Van Winsum. The functional visual field as indicator of workload. *Proceedings of the meeting of the Dutch ergonomics association*, pages 182–189, 1999.

[228] Y. Kume, A. Shirai, M. Tsuda, and T. Hatada. Information transmission through soles by vibrotactile stimulation. In *Transaction of the Virtual Reality Society of Japan*, volume 3, pages 83–88, 1998.

[229] Martin Frey. CabBoots: Shoes with integrated guidance system. In *Proceedings of the 1st international conference on tangible and embedded interaction (TEI'07)*, pages 245–246, New York, NY, USA, 2007. ACM.

[230] Hyunho Kim, Changhoon Seo, Junhun Lee, Jeha Ryu, Si bok Yu, and Sooyoung Lee. Vibrotactile display for driving safety information. *IEEE Intelligent Transportation Systems Conference (ITSC'06)*, pages 573–577, 2006.

[231] Michael W. Burke, Richard D. Gilson, and Richard J. Jagacinski. Multi-modal information processing for visual workload relief. *Ergonomics, Taylor and Francis Ltd.*, 23(10):961–975, 1980.

[232] Tomohiro Amemiya, Hideyuki Ando, and Taro Maeda. Phantom-DRAWN: direction guidance using rapid and asymmetric acceleration weighted by nonlinearity of perception. In *Proceedings of the 2005 international conference on augmented tele-existence (ICAT'05)*, pages 201–208, New York, NY, USA, 2005. ACM.

[233] J. Schumann, H. Godthelp, B. Farber, H. Wontorra, A. Gale, I. Brown, C. Haslegrave, H. Kruysse, and S. Taylor. Breaking up Open-Loop Steering Control Actions: The Steering Wheel as an Active Control Device. In A. G. Gale, I. D. Brown, C. M. Haslegrave, H. W. Kruysse, and S. P. Taylor, editors, *Proceedings of the Vision in Vehicles*, volume 4, pages 321–332, Amsterdam, 1993. Elsevier (Sciencedirect).

[234] H.S. Vitense, J.A. Jacko, and V.K. Emery. Multimodal feedback: an assessment of performance and mental workload. *Ergonomics, Taylor and Francis Ltd.*, 46(1–3):68–87, January 2003.

[235] Holly S. Vitense, Julie A. Jacko, and V. Kathlene Emery. Multimodal feedback: establishing a performance baseline for improved access by individuals with visual impairments. In *Proceedings of the fifth international ACM conference on assistive technologies (Assets'02)*, pages 49–56, New York, NY, USA, July 8–10 2002. ACM. ISBN: 1-58113-464-9.

[236] John D. Lee, Joshua D. Hoffman, and Elizabeth Hayes. Collision warning design to mitigate driver distraction. In *Proceedings of the SIGCHI conference on human factors in computing systems (CHI'04)*, pages 65–72, New York, NY, USA, 2004. ACM.

[237] Aaron E. Sklar and Nadine B. Sarter. Good Vibrations: Tactile Feedback in Support of Attention Allocation and Human-Automation Coordination in Event-Driven Domains. *Human Factors*, 41(4):543–552, 1999.

[238] Luv Kohli, Masataka Niwa, Haruo Noma, Kenji Susami, Kenichi Hosaka, Yasuyuki Yanagida, Robert W. Lindeman, and Yuichiro Kume. Towards Effective Information Display Using Vibrotactile Apparent Motion. In *Proceedings of the Symposium on Haptic Interfaces for Virtual Environment and Teleoperator Systems (HAPTICS'06)*, pages 445–451, Washington, DC, USA, 2006. IEEE Computer Society.

[239] P. A. Fenety, C. Putnam, and J. M. Walker. In-chair movement: validity, reliability and implications for measuring sitting discomfort. *Applied Ergonomics*, 31, 2000.

[240] L. A. Slivosky and H. Z. Tan. A Real-Time Static Posture Classification System. In S. Nair, editor, *Proceedings on 9th International Symposium on Haptic Interfaces for Virtual Environment and Teleoperator Systems*, volume 69, pages 1049–1056. American Society of Mechanical Engineers Dynamic Systems and Control Division, Orlando, 2000.

[241] Lynne A. Slivosky and Hong Z. Tan. A Real-Time Sitting Posture Tracking System. ECE Technical Reports TR-ECE 00-1, Purdue University School of ECE (Electrical and Computer Engineering), School of Electrical and Computer Engineering, 1285 Electrical Engineering Building, Purdue University, West Lafayette, Indiana, 47907–1285, January 2001.

[242] K. J. Overbeeke, P. Vink, and F. K. Cheung. The emotion-aware office chair. In *Proceedings of International Conference on Affective Human Factors Design*. Asean Academic Press, London, 2001.

[243] Selena Mota and Rosalind W.Picard. Automated Posture Analysis for Detecting Learner's Interest Level. In *Workshop on Computer Vision and Pattern Recognition for Human-Computer Interaction (CVPR HCI '03)*, 2003.

[244] Giuseppe Andreoni, Giorgio C. Santambrogio, Marco Rabuffetti, and Antionio Pedotti. Method for the analysis of posture and interface pressure of car drivers. volume 33, pages 511 – 522. Department of Bioengineering, Polytechnic of Milan, Plazza Leodnardo da Vinci, Milan, Italy and Centro di Bioingegnerio, Via Capecelatro, Milan, Italy, Elsievier Science Ltd., Applied Ergonomics, 2002.

[245] Hermann Miller. Body support in the office: Sitting, seating, and low back pain. Technical report, Hermann Miller Inc., 2002.

[246] Jennifer Healey and Rosalind Picard. SmartCar: detecting driver stress. In *Proceedings of 15th International Conference on Pattern Recognition (ICPR 2000)*, volume 4, pages 218–221 (4), Barcelona, Spain, September 2000. IEEE Computer Society Press.

[247] Manli Zhu, Alex M. Martinez, and Hong Z. Tan. Template-based recognition of static sitting postures. In *Proceedings of the 2003 Conference on Computer Vision and Pattern Recognition Workshop (CVPRW'03)*. Dept. of Electrical Engineering, The Ohio State University and School of Electrical and Computer Engineering, Purdue University, IEEE Computer Society Press, 2003.

[248] Rui Ishiyama, Hiroo Ikeda, and Shizuo Sakamoto. A compact model of human postures extracting common motion from individual samples. In *Proceedings of the 18th International Conference on Pattern Recognition (ICPR'O6)*, page 4. Media and Information Research Laboratories, NEC Corporation, IEEE Computer Society Press, 2006.

[249] Benjamin A.C. Forsyth and Karon E. MacLean. Predictive haptic guidance: Intelligent user assistance for the control of dynamic tasks. *IEEE Transactions on Visualization and Computer Graphics*, 12(1):103–113, 2006.

[250] Stephen Brewster, Faraz Chohan, and Lorna Brown. Tactile feedback for mobile interactions. In *Proceedings of the SIGCHI conference on Human factors in computing systems (CHI '07)*, pages 159–162, New York, NY, USA, 2007. ACM Press.

[251] Chee Fai Tan, Frank Delbressine, and Matthias Rauterberg. Vehicle seat design: state of the art and recent development. In *Proceedings of World Engineering Congress (WEC07), Penang, Malaysia, 5–9 August*, page 12, 2007.

[252] Junhun Lee, Hyunho Kim, Ryu, J., and Jinmyoung Woo. Preliminary User Studies on Tactile Interface for Telematics Services. *IEEE Intelligent Transportation Systems Conference (ITSC 2007)*, pages 929–933, September 30 – October 3, 2007. ISBN: 978-1-4244-1396-6.

[253] Dorit Haider. *Der Renault Laguna*. Renault Presse & Öffentlichkeitsarbeit, Renault Österreich GmbH, Laaer Berg-Strasse 64, A-1101 Vienna, November 2007. Online, last retrieved August 29, 2009. http://www.media.renault.at/__/?3359.013e01bb.DL.

[254] John McCormick. Crash Avoidance Goes High-Tech. *Automotive Industries*, 184(11):36–39, November 2004.

[255] Johan Noren. Warning systems design in a glass cockpit environment. Liu-iei-tek-a–08/00335–se, Department for Management and Engineering, University of Linköping, January 2008.

[256] Akua B. Ofori-Boateng. A Study of the Effect of Varying Air-Inflated Seat Cushion Parameters on Seating Comfort. Master thesis, Virginia Polytechnic Institute and State University, Blacksburg, Virginia, June 2003.

[257] M. J. Griffin, H. V. Howarth, P. M. Pitts, S .Fischer, U. Kaulbars, P. M. Donati, and P. F. Bereton. Guide to good practice on Whole-Body Vibration. Practice Guide V6.7g, Advisory Committee on Safety and Health at Work, European Commission, June 12 2006.

[258] Marianne Schust, Ralph Blüthner, and Helmut Seidel. Examination of perceptions (intensity, seat comfort, effort) and reaction times (brake and accelerator) during low-frequency vibration in x- or y-direction and biaxial (xy-) vibration of driver seats with activated and deactivated suspension. *Journal of Sound and Vibration*, 298(3):606–626, December 2006.

[259] J. L. Van Niekerk, W. J. Pielemeier, and J. A. Greenberg. The use of seat effective amplitude transmissibility (seat) values to predict dynamic seat comfort. *Journal of Sound and Vibration*, 260(5):867–888, March 2003.

[260] Jian-Da Wu and Rong-Jun Chen. Application of an active controller for reducing small-amplitude vertical vibration in a vehicle seat. *Journal of Sound and Vibration*, 274(3-5):939–951, July 2004.

[261] Wassim El Falou, Jacques Duchêne, Michel Grabisch, David Hewson, Yves Langeron, and Frédéric Lino. Evaluation of driver discomfort during long-duration car driving. *Applied Ergonomics*, 34(3):249–255, May 2003.

[262] G. S. Paddan and J. M. Griffin. Effect Of Seating On Exposures To Whole-Body Vibration In Vehicles. *Journal of Sound and Vibration*, 253(1):215–241, May 2002.

[263] H. K. Jang and M. J. Griffin. Effect of phase, frequency, magnitude and posture on discomfort associated with differential vertical vibration at the seat and feet. *Journal of Sound and Vibration*, 229(2):273–286, January 2000.

[264] M. Demic, J. Lukic, and Z. Milic. Some aspects of the investigation of random vibration influence on ride comfort. *Journal of Sound and Vibration*, 253(1):109–128, May 2002.

[265] Andras Varhelyi. Speed management via in-car devices: effects, implications, perspectives. *Transportation*, 29(3):237–252, August 2002.

[266] Heyms et al. CAR 2 CAR Communication Consortium Manifesto: Overview of the C2C-CC System, Version 1.1. Technical report, August 28, 2007.

[267] M.M. Frechin, S.B. Arino, and J. Fontaine. ACTISEAT: active vehicle seat for acceleration compensation. *Proceedings of the I MECH E Part D Journal of Automobile Engineering*, 218:925–933, September 1 2004.

[268] Alberto Rovetta, Chiara Zocchi, Alessandro Giusti, Alessandro Adami, and Francesco Scaramellini. Haptics and biorobotics for increasing automotive safety. Submitted to IEEE International Conference on Intelligent Robots and Systems (IROS'06), March 2006.

[269] J. R. Treat, N.S. Tumbas, S.T. McDonald, D. Shinar, R.D. Hume, R.E. Mayer, R.L. Stanisfer, and N.J. Castellan. Tri-Level Study of the Causes of Traffic Accidents. Volume 1: Causal Factor Tabulations and Assessments. Interim Report DOT-HS-805-085, US Department of Transportation, National Highway Traffic Safety Administration, Washington D.C., 1979.

[270] Douglas J. Beirness, Herb M. Simpson, and Katharine Desmond. The Road Safety Monitor 2002: Risky Driving. Annual Survey, Traffic Injury Research Foundation (TIRF), 171 Nepean Street, Suite 200, Ottawa, Ontario K2P 0B4, November 2002. ISBN: 0-920071-27-9.

[271] Ing-Marie Jonsson and Fang Chen. In-vehicle information system used in complex and low traffic situations: Impact on driving performance and attitude. In *HCI (6)*, pages 421–430, 2007.

[272] C. Baber, B. Mellor, R. Graham, J.M. Noyes, and C Tunley. Workload and the use of automatic speech recognition: The effects of time and resource demands. *Speech Communication (Speech under Stress)*, 20(1–2):37–53 (17), November 1996.

[273] Dominique Freard, Eric Jamet, Olivier Le Bohec, Gerard Poulain, and Valerie Botherel. Subjective Measurement of Workload Related to a Multimodal Interaction Task: NASA-TLX vs. Workload Profile. In J. Jacko, editor, *12th International Conference on Human-Computer Interaction (HCII'07), Beijing, China*, LNCS 4552, pages 60–69. Springer-Verlag Berlin Heidelberg, July 2007.

[274] Fred Paas and Jeroen Van Merrienboer. Instructional control of cognitive load in the training of complex cognitive tasks. *Educational Psychology Review*, 6(4):351–371, December 1994.

[275] Fred Paas and Jeroen Van Merrienboer. Measurment of cognitive load in instructional research. *Perceptual and Motor Skills*, 79(4):419–430, December 1994.

[276] John Sweller, Jeroen Van Merrienboer, and Fred Paas. Cognitive Architecture and Instructional Design. *Educational Psychology Review*, 10(3):251–296, September 1998.

[277] Joanne L. Harbluk, Y. Ian Noy, and Moshe Eizenman. The Impact of Cognitive Distraction on Driver Visual Behaviour and Vehicle Control. TP 13889E, Transport Canada, Ergonomics Division, Road Safety Directorate and Motor Vehicle Regulation Directorate, February 2002.

[278] Dave Lamble, Tatu Kauranen, Matti Laakso, and Heikki Summala. Cognitive load and detection thresholds in car following situations: Safety implications for using mobile (cellular) telephones while driving. *Accident Analysis & Prevention*, 31(6):617–623, November 1999.

[279] Joshua B. Hurwitz and David J. Wheatley. Driver Choice of Headway with Auditory Warnings. *Human Factors and Ergonomics Society Annual Meeting Proceedings*, 45:1637–1640(4), 2001.

[280] John D. Lee, Daniel V. McGehee, Timothy L. Brown, and Michelle L. Reyes. Collision Warning Timing, Driver Distraction, and Driver Response to Imminent Rear-End Collisions in a High-Fidelity Driving Simulator. *Human Factors: The Journal of the Human Factors and Ergonomics Society*, 44:314–334(21), 2002.

[281] Chip Wood and Joshua Hurwitz. Driver workload management during cell phone conversations. In *Proceedings of the 3rd International Driving Symposium on Human Factors in Driver Assessment, Training, and Vehicle Design*, pages 2022 – 209, June 27–30, 2005.

[282] Peter J. Cooper, Yvonne Zheng, Christian Richard, John Vavrik, Brad Heinrichs, and Gunter Siegmund. The impact of hands-free message reception/response on driving task performance. *Accident Analysis & Prevention*, 35(1):23–35, January 2003.

[283] Mark A. Neerincx. *Handbook of Cognitive Task Design*, chapter Cognitive Task Load Analysis: Allocating Tasks and Designing Support, pages 283–306. Lawrence Erlbaum Associates, 2003. ISBN: 0-805-84003-6.

[284] R. Broström, J. Engström, A. Agnvall, and G. Markkula. Towards the next generation intelligent driver information system (IDIS): The volvo cars interaction manager concept. In *Proceedings of the 2006 Intelligent Transport Systems World Congress, London*, 2006.

[285] P. Green. Driver distraction, telematics design, and workload managers: Safety issues and solutions. *SAE Paper Number 2004-21-0022*, 2004. 16 pages.

[286] Johan Engstrom and Louise Floberg. Method and device for estimating workload for a driver of a vehicle. (European Patent EP1512584), March 2008. Lilliegatan 10, SE-416 57 Göteborg, Sweden.

[287] Sean Peirce and Jane Lappin. Private Sector Deployment of Intelligent Transportation Systems: Current Status and Trends. Final, 2004–2006 HW52/ CK288, US Department of Transportation, Research and Innovative Technology Administration, John A. Volpe National Transportation Systems Center, 55 Broadway, Cambridge, MA 02142, February 2006. 34 pages.

[288] Frédéric Vernier and Laurence Nigay. A framework for the combination and characterization of output modalities. In *DSV-IS*, pages 35–50, 2000.

[289] J. R. Treat. A study of pre-crash factors involved in traffic accidents. *HSRI Research Review 10(6)/11(1)*, pages 1–35, 1980.

[290] Laurence R. Young, Kathleen H. Sienko, Lisette E. Lyne, Heiko Hecht, and Alan Natapoff. Adaptation of the vestibulo-ocular reflex, subjective tilt, and motion sickness to head movements during short-radius centrifugation. *Journal of Vestibular Research*, 13(2):65–77, January 2003.

[291] Dawn Royal. National Survey of Distracted and Drowsy Driving Attitudes and Behavior: 2002, Volume I: Findings. Survey DOT HS 809 566, National Highway Traffic Safety Administration, Washington D.C., The Gallup Organization, 10. Work Unit No. (TRAIS), 901 F Street, NW - Suite 400, Washington DC 20004 (Performing Organization), April 2003. 68 pages.

[292] Jane C. Stutts, Donald W. Reinfurt, Loren Staplin, and Eric A. Rodgman. The Role of Driver Distraction in Traffic Crashes. Technical report, University of North Carolina, Highway Safety Research Center, Chapel Hill, NC, prepared for AAA Foundation for

Traffic Safety, 1440 New York Avenue, N.W., Suite 201, Washington, DC 20005, May 2001. ISBN 0-309-08760-0.

[293] Oliver Paczkowski and Jörg Müller. Identification of Driver Distraction in Automotive Environments. In *2nd Workshop on Automotive User Interfaces and Interactive Applications (AUIIA'08), Mensch und Computer 2008, Luebeck, Germany*. Logos Verlag, Berlin, September 7–10, 2008. ISBN: 978-3-8325-2007-6, 6 pages.

[294] K. Torkkola, N. Massey, and C. Wood. Detecting driver inattention in the absence of driver monitoring sensors. *Machine Learning and Applications, 2004. Proceedings. 2004 International Conference on*, pages 220–226, December 2004. ISBN: 0-7803-8823-2.

[295] Jane Stutts, Ronald R. Knipling, Ronald Pfefer, Timothy R. Neuman, Kevin L. Slack, and Kelly K. Hardy. Guidance for Implementation of the AASHTO Strategic Highway Safety Plan. Volume 14: A Guide for Reducing Crashes Involving Drowsy and Distracted Drivers. Technical Report Project G17-18(3), NHCRP Report 500, National Cooperative Highway Research Program, Transportation Research Board, Business Office, 500 Fifth Street, NW, Washington, DC 20001, 2005. ISBN 0-309-08760-0.

[296] Richard Parry-Jones. Vehicle Infotronics: Enabling the Integrated Mobility Experience. In *Proceedings of the 1998 International Congress on Transportation Electronics (Convergence'98)*. Society of Automotive Engineers (SAE), October 19–21 1998. ISBN-13: 978-0-768-00277-5, 444 pages.

[297] Tara Matthews, Anind K. Dey, Jennifer Mankoff, Scott Carter, and Tye Rattenbury. A toolkit for managing user attention in peripheral displays. In *Proceedings of the 17th annual ACM symposium on user interface software and technology (UIST'04)*, pages 247–256, New York, NY, USA, 2004. ACM.

[298] V. Laurutis. Channel Information Capacity of the Sensomotor System of the Eye. *Electronics and Electrical Engineering*, 85(5):85–88, 2008. ISSN: 1392-1215.

[299] Claude E. Shannon. A mathematical theory of communication. *Bell System Technical Journal*, 27:379–423, July 1948.

[300] R.L. French. In-vehicle navigation–status and safety impacts. In *Technical Papers from ITE's 1990, 1989, and 1988 Conference*. Institute of Transportation Engineers, Washington, DC, 1990. pages 226–235.

[301] G. Labiale. Influence of in-car navigation map displays on driver performances. SAE Technical Paper Series 891683, Society of Automotive Engineers, Warrendale, PA, 1989.

[302] T. A. Dingus and M. C Hulse. Some Human Factors Design Issues and Recommendations for Automobile Navigation Information Systems. Transportation Research C 1,2, 1993. page 119 – 131.

[303] Joachim Kaufmann. Head-up-Display: Neue Technik für mehr Verkehrssicherheit (published October 20, 2004). Online, last retrieved August 15, 2009. http://www.zdnet.de/enterprise/tech/auto/0,39026506,39125753-1,00.htm.

[304] Dean Pomerleau, Todd Jochem, Charles Thorpe, Parag Batavia, Doug Pape, Jeff Hadden, Nancy McMillan, Nathan Brown, and Jeff Everson. Run-Off-Road Collision Avoidance Countermeasures Using IHVS Countermeasures. Final Report DOT HS 809 170, National Highway Traffic Safety Administration (NHTSA), 400 Seventh Street, S.W., Washington, DC 20590, December 1999. 134 pages.

[305] T. C. Lansdown. Driver visual allocation and the introduction of intelligent transport systems. *Proceedings of the Institution of Mechanical Engineers, Part D: Journal of Automobile Engineering*, 214(6):645–652, January 2000.

[306] Paul Green, William Levison, Gretchen Paelke, and Colleen Serafin. Preliminary Human Factors Design Guidelines for Driver Information Systems. Technical Report FHWA-RD-94-087, Federal Highway Administration, Office of Safety and Traffic Operations R&D, 6300 Georgetown Pike, McLean, Virginia 221 01 -2296, December 1995. 134 pages.

[307] J. Walker, E. Alicandri, C. Sedney, and K. Roberts. In-vehicle navigation devices: Effects on the safety of driver performance. *Vehicle Navigation and Information Systems Conference, 1991*, 2:499–525, October 1991.

[308] Bobbie Seppelt and Christopher D. Wickens. In-Vehicle Tasks: Effects of Modality, Driving Relevance, and Redundancy. Technical Report AHFD-03-16/GM-03-2, Aviation Human Factors Division, Institute of Aviation, University of Illinois at Urbana-Champaign, Savoy, Illinois 61874, August 2003.

[309] James L. Szalma, Joel S. Warm, Gerald Matthews, William N. Dember, Ernest M. Weiler, Ashley Meier, and Thomas F. Eggemeier. Effects of sensory modality and task duration on performance, workload, and stress in sustained attention. *Human Factors*, June 2004.

[310] Mark Vollrath and Ingo Totzke. In-vehicle communication and driving: an attempt to overcome their interference. Center for Traffic Sciences (IZVW), University of Wuerzburg, Germany, sponsored by the United States Department of Transportation, 2000.

[311] Emily E. Wiese and John D. Lee. Auditory alerts for in-vehicle information systems: The effects of temporal conflict and sound parameters on driver attitudes and performance. *Ergonomics*, 47(9):965–986 (22), July 2004.

[312] Chad Brooks and Andry Rakotonirainy. In-vehicle technologies, Advanced Driving Assistance Systems and driver distraction: Research challenges. In *Proceedings of the 1st International ARCS Conference on Distracted Driving, Sydney, Australia*. Centre for Accident Research and Road Safety, Queensland, Australia, June 2005.

[313] T. Philips. ITS World Congress 2008: Mercedes-Benz to Present the Driver Assistance Systems of Tomorrow. Online, last retrieved August 31, 2009 (published online November 17, 2008). http://www.emercedesbenz.com/Nov08/17_001514_Mercedes_Benz_To_Present_Driver_Assistance_Systems_Of_Tomorrow_At_The_ITS_World_Congress_2008.html.

[314] A. Weithöner. Driver Assistance Systems. Technical Information, Hella KGaA Hueck & Co., Rixbecker Straße 75, D-59552 Lippstadt/Germany, November 2007. Online, last retrieved August 31, 2009. http://www.hella.com/hella-com-en/assets/media_global/AutoIndustrie_ti_fas_gb.pdf.

[315] Günther Nirschl. Human-Centered Development of Advanced Driver Assistance Systems. *Human Interface and the Management of Information. Interacting in Information Environments*, pages 1088–1097, August 2007.

[316] Jürgen Goroncy. Sensors See the Light. *Siemens Pictures of the Future*, pages 47–49, Fall 2005.

[317] Karel A. Brookhuis, Dick de Waard, and Wiel H. Janssen. Behavioural Impacts of Advanced Driver Assistance Systems – an Overview. *European Journal of Transport and Infrastructure Research (EJTIR), Special Issue: Implementation Issues of Advanced Driver Assistance Systems*, 3:245–254, November 2001. J. Marchau and K.A. Brookhuis (Eds.).

[318] ICT Results. Driving the future of in-vehicle ICT. Online, last retrieved August 29, 2009. http://www.physorg.com/news139831296.html.

[319] ICT Results. Dashing computer interface to control your car. Fact sheet INF70100, Project Adaptive integrated driver-vehicle interface (AIDE), August 2008. Online, last retrieved August 31, 2009. http://www.aide-eu.org.

[320] F. Tango and R. Montanari. In-car machine-human interaction: how the new vehicle technologies which respond to the vehicle's needs could match with the user-centered

approach and contribute to shape a user-centered design approach. *IEEE International Conference on Systems, Man and Cybernetics*, 3:2558 – 2563 vol.3, October 2004. ISBN: 0-7803-8566-7.

[321] Immersion. Immersion: The Value of Haptics. Technical report, Immersion Corporation, 801 Fox Lane, San Jose, CA 95131 USA, May 2007.

[322] K. Suzuki and H. Jansson. An analysis of driver's steering behaviour during auditory or haptic warnings for the designing of lane departure warning system. *JSAE Review*, 24(1):65–70(6), January 2003.

[323] Motoyuki Akamatsu, Sigeru Sato, and I. Scott MacKenzie. Multimodal mouse: A mouse-type device with tactile and force display. *Presence*, 3(1):73–80, 1994.

[324] Thomas Debus, Theresia Becker, Pierre Dupont, Tae-Jeong Jang, and Robert Howe. Multichannel vibrotactile display for sensory substitution during teleoperation. In *Proceedings of SPIE – The International Society for Optical Engineering*, volume 4570, pages 42–49, 2001.

[325] A. Bicchi, E.P. Scilingo, D. Dente, and N. Sgambelluri. Tactile Flow and Haptic Discrimination of Softness. In F. Barbagli, D. Prattichizzo, and K. Salisbury, editors, *Multi-point Interaction with Real and Virtual Objects*, number 18 in Springer Tracts in Advanced Robotics (STAR), pages 165–176. Springer Verlag, 2005.

[326] Wolfang Wünschmann and D. Fourney. Guidance on Tactile Human-System Interaction: Some Statements. In *Proc. Workshop GOTHI'05 (Organized by ISO TC 159/SC4/WG9)*, Saskatoom, Saskatchewan, Canada, 2005.

[327] Julia Layton. How BrainPort Works: Concepts of Electrotactile Stimulation. Online, last retrieved August 27, 2009 (published July 17, 2006). http://science.howstuffworks.com/brainport1.htm.

[328] Lynette A. Jones and Michal Berris. The Psychophysics of Temperature Perception and Thermal-Interface Design. In *HAPTICS '02: Proceedings of the 10th Symposium on Haptic Interfaces for Virtual Environment and Teleoperator Systems*, page 137, Washington, DC, USA, 2002. IEEE Computer Society.

[329] Kenichi Amemiya and Yutaka Tanaka. Portable Tactile Feedback Interface Using Air Jet. In *Proceedings of the 9th International Conference on Artificial Reality and Teleexistence*, 1999. 18 pages.

[330] Robert Stone. Haptic feedback: a brief history from telepresence to virtual reality. *Haptic Human-Computer Interaction*, pages 1–16, 2001.

[331] Disc translator HVPZT, types P-286, P-288, P-289. Technical datasheet, Physik Instrumente (PI) GmbH & Co. KG, Auf der Römerstrasse 1, D-76228 Karlsruhe, November 2006. http://www.physikinstrumente.de/de/pdf/P286_Datenblatt.pdf.

[332] Dimitrios A. Kontarinis and Robert D. Howe. Display of high-frequency tactile information to teleoperators. volume 2057, pages 40–50. SPIE, 1993.

[333] K.B. Shimoga. A survey of perceptual feedback issues in dexterous telemanipulation: Part ii. finger touch feedback. In *IEEE Virtual Reality Annual International Symposium (VRAIS '93), New York*, pages 271–279, September 1993.

[334] S. J. Bolanowski, Jr., G. A. Gescheider, R. T. Verrillo, and C. M. Checkosky. Four channels mediate the mechanical aspects of touch. *Acoustical Society of America Journal*, 84:1680–1694, November 1988.

[335] F. Reynier and V. Hayward. Summary of the kinesthetic and tactile function of the human upper extremity. Technical Report CIM-93-4, McGill Research Center for Intelligent Machines, McGill University, Montreal, Quebec, Canada, 3480 University St., Montreal, Quebec, Canada, H3A 2A7, March 1993.

[336] Wikipedia. Mechanoreceptor. Online, last retrieved August 30, 2009. http://en.wikipedia.org/wiki/Mechanoreceptor.

[337] José Ochoa and Erik Torebjörk. Sensations evoked by intraneural microstimulation of single mechanoreceptor units innervating the human hand. *The Journal of Physiology*, 342(1):633–654, 1983.

[338] Heidi Wikström. *Neuromagnetic Studies on Somatosensory Functions in Healthy Subjects and Stroke Patients*. PhD thesis, Department of Clinical Neurosciences, University of Helsinki, December 1999. ISBN: 951-45-9004-X (PDF version).

[339] Ingrid Moll, Marion Roessler, Johanna M. Brandner, Ann-Christin Eispert, Pia Houdek, and Roland Moll. Human Merkel cells – aspects of cell biology, distribution and functions. *European Journal of Cell Biology*, 84(2–3):259–271, 2005.

[340] A. Bicchi, D. Dente, and E. P. Scilingo. Haptic Illusions Induced by Tactile Flow. In *EuroHaptics Conference*, pages 2412–2417, 2003.

[341] Engineering Acoustics, Inc. (EAI). Physiology of Haptic Perception. Online, last retrieved August 30, 2009. http://www.eaiinfo.com//ProductData/TactorPhysiology.htm.

[342] K. Bark, J. W. Wheeler, S. Premakumar, and M. R. Cutkosky. Comparison of Skin Stretch and Vibrotactile Stimulation for Feedback of Proprioceptive Information. *Symposium on haptic interfaces for virtual environment and teleoperator systems*, pages 71–78, March 2008.

[343] Roger W. Cholewiak, J. Christopher Brill, and Anja Schwab. Vibrotactile Localization on the Abdomen: Effects of Place and Space. *Perception & Psychophysics*, 66(6):970–987, August 2004.

[344] Lorna Margaret Brown. *Tactons: Structured Vibrotactile Messages for Non-Visual Information Display*. PhD thesis, Department of Computing Science, University of Glasgow, April 2007.

[345] Kimberly Myles and Mary S. Binseel. The Tactile Modality: A Review of Tactile Sensitivity and Human Tactile Interfaces. Technical Report ARL-TR-4115, U.S. Army Research Laboratory (ARL), Human Research and Engineering Directorate, Aberdeen Proving Ground, MD 21005-5425, May 2007.

[346] Mario J. Enriquez, Karon E. MacLean, and Christian Chita. Haptic Phonemes: Basic Building Blocks of Haptic Communication. In Francis K. H. Quek, Jie Yang, Dominic W. Massaro, Abeer A. Alwan, and Timothy J. Hazen, editors, *ICMI*, pages 302–309. ACM, 2006.

[347] Vincent Hayward, Oliver R. Astley, Manuel Cruz-Hernandez, Danny Grant, and Gabriel Robles-De-La-Torre. Haptic Interfaces and Devices. *Sensor Review*, 24(1):16–29, 2004.

[348] Joseph Gaspard. Hearing and tactile detection in manatees. Online, last retrieved August 27, 2009. http://www.marine.usf.edu/bio/fishlab/research/research2.htm.

[349] Hannu Virokannas. Vibration perception thresholds in workers exposed to vibration. *International Archives of Occupational and Environmental Health*, 64(5):377–382, December 1992. ISSN: 0340-0131 (Print), 1432-1246 (Online).

[350] G. Lundborg, A. K. Lie-Stenström, C. Sollerman, T. Strömberg, and I. Pyykkö. Digital vibrogram: a new diagnostic tool for sensory testing in compression neuropathy. In *The journal of hand surgery*, volume 11, pages 693–699, September 1986.

[351] A. J. Brammer, J. E. Piercy, S. Nohara, H. Nakamura, P. L. Auger, A. T. Haines, M. Lawrence, R. L. Brubaker, and C. van Netten. Vibrotactile thresholds in operators of vibrating hand-held tools. In A. Akada, W. Taylor, and H. Dupuis, editors, *International Conference on Hand-Arm Vibration*, pages 221–223. Kyoei Press, Kanazawa Japan, 1990.

[352] Advanced Fair, Rue du Collège St. Michel 9-11 B-1150 1150 Brussels, Belgium ECR (Efficient Consumer Response) Europe. The moment of truth – Putting category management into action – At store level. Powerpoint presentation, Online, last retrieved August 25, 2009. http://forum.ecrnet.org/PublicPages/Archive/Berlin/Downloads/CC1-CategoryManagement.pdf.

[353] Advanced Fair, Rue du Collège St. Michel 9-11 B-1150 1150 Brussels, Belgium ECR (Efficient Consumer Response) Europe. ECR Europe Category Management Best Practices Report: Chapter 4.7 Category Plan Implementation. Report, Online, last retrieved August 26, 2009. http://www.ecrnet.org/demand_side/demand_side_catman_bluebook.html.

[354] Anthony J. Brammer and Joseph E. Piercy. Method and system for identifying vibrotactile perception thresholds of nerve endings with inconsistent subject response rejection. (US Patent 5433211), July 18, 1995. National Research Council of Canada, Ottawa, Canada.

[355] Charles J. Testa and Douglas B. Dearie. Human factors design criteria in man-computer interaction. In *ACM 74: Proceedings of the 1974 annual conference*, pages 61–65, New York, NY, USA, 1974. ACM.

[356] R. B. Miller. Response time in man-computer conversational transactions. In *Proceedings of AFIPS Fall Joint Computer Conference*, volume 33, pages 267–277, 1968.

[357] Ben Shneiderman. Response time and display rate in human performance with computers. *ACM Computing Surveys*, 16(3):265–285, September 1984.

[358] Jeffrey J. Scott and Rob Gray. Comparison of Driver Brake Reaction Times to Multimodal Rear-End Collision Warnings. In *Proceedings of the Fourth International Driving Symposium on Human Factors in Driver Assessment, Training and Vehicle Design, Stevenson, Washington USA*. Arizona State University and United States Air Force, Mesa, AZ, USA, July 2007.

[359] H. Kajimoto, N. Kawakami, and S. Tachi. Psychophysical evaluation of receptor selectivity in electro-tactile display. In *13th International Symposium on Measurement and Control in Robotics (ISMCR'03), Madrid, Spain*, December 2003. 4 pages.

[360] Yon Visell. Tactile sensory substitution: Models for enaction in HCI. *Interacting with Computers*, August 2008. 16 pages.

[361] Helena Pongrac. Vibrotactile perception: examining the coding of vibrations and the just noticeable difference under various conditions. *Multimedia Systems*, 13(4):297–307, January 2008.

[362] Lynette Jones. Human Factors Overview. In *Haptics, Virtual Reality, and Human Computer Interaction*. Massachusetts Institute of Technology, June 2001.

[363] K. A. Kaczmarek, J. G. Webster, P. Bach y Rita, and W. J. Tompkins. Electrotactile and vibrotactile displays for sensory substitution systems. *IEEE Transactions on Biomedical Engineering*, 38(1):1–16, January 1991. ISSN: 0018-9294.

[364] G. A. Gescheider and R. T. Verrillo. *Sensory functions of the skin of humans*, chapter Vibrotactile frequency characteristics as determined by adaptation and masking procedures, pages 183–203. Plenum Press, New York, 1979. ISBN: 0-30-640321-8.

[365] R. T. Verrillo. Temporal Summation in Vibrotactile Sensitivity. *The Journal of the Acoustical Society of America*, 37(5), 1965.

[366] R. W. Cholewiak and C. McGrath. Vibrotactile Targeting in Multimodal Systems: Accuracy and Interaction. *14th Symposium on Haptic Interfaces for Virtual Environment and Teleoperator Systems*, pages 413–420, March 2006.

[367] Gregory M. Fitch, Raymond J. Kiefer, Jonathan M. Hankey, and Brian M. Kleiner. Toward Developing an Approach for Alerting Drivers to the Direction of a Crash Threat. *Human Factors: The Journal of the Human Factors and Ergonomics Society*, 49:710–720(11), August 2007.

[368] Andrea S. Krausman and Timothy L. White. Tactile Displays and Detectability of Vibrotactile Patterns as Combat Assault Maneuvers are Being Performed. Technical Report ARL-TR-3998, U.S. Army Research Laboratory (ARL), Human Research and Engineering Directorate, Aberdeen Proving Ground, MD 21005-5425, December 2006.

[369] Zachary Pousman and John Stasko. A taxonomy of ambient information systems: four patterns of design. In *Proceedings of the working conference on advanced visual interfaces (AVI'06)*, pages 67–74, New York, NY, USA, 2006. ACM.

[370] Jan Hermkens. *Tools for Professionals: FSA documentation*. Vista Medical Europe B.V., Industrieterrein 40, NL-5981 AK Panningen, The Netherlands, August 2, 2006.

[371] James T. Fulton. Light & dark adaptation in human vision, November 2005.

[372] Linday I. Smith. A Tutorial on Principal Components Analysis, 26 pages. February 2002.

[373] K. Fukunaga. *Introduction to Statistical Pattern Recognition (Second Edition)*. Academic Press, New York, 1990.

[374] Aapo Hyvärinen, Juha Karhunen, and Erkki Oja. *Independent Component Analysis*. John Wiley & Sons, Inc. 605 Third Avenue, New York, 2001. ISBN: 0-471-40540-X.

[375] Ian T. Jolliffe. *Principal component analysis*. Springer series in statistics. Springer-Verlag New York, Inc., 2nd edition, 2002. ISBN: 0-387-95442-2.

[376] Kimberly Patch. Biometrics takes a seat. Online, last retrieved October 11, 2009 (published online November 15, 2000). Technology Research News, http://www.trnmag.com/Stories/111500/Seat_Tracking_111500.html.

[377] R. A. Fisher. The use of multiple measurements in taxonomic problems. *Annals of Eugenics*, 7:179–188, 1936.

[378] Geoffrey J. McLachlan. *Discriminant Analysis and Statistical Pattern Recognition*. Wiley-Interscience, August 4, 2004. ISBN: 04-716-9115-1, 552 pages.

[379] R.O. Duda, P.E. Hart, and D.H. Stork. *Pattern Classification*. Wiley Interscience, 2nd edition, 2000. ISBN: 0-471-05669-3.

[380] Jerome H. Friedman. Regularized discriminant analysis. In *Journal of the American Statistical Association*. Department of Statistics and Stanford Linear Accelerator Center, Stanford University, Stanford CA 94309, 1999.

[381] A. M. Martinez and A. C. Kak. PCA versus LDA. In *IEEE Transactions on Pattern Analysis and Machine Intelligence*, volume 23, pages 228–233, 2001.

[382] Nagendra Kumar Goe. *Investigation of Silicon Auditory Models and Generalization of Linear Discriminant Analysis for Improved Speech Recognition (Chapter 5)*. PhD thesis, Johns Hopkins University, 1997.

[383] S. Balakrishnama and A. Ganapathiraju. Linear discriminant analysis - a brief tutorial. Technical report, Institute for Signal and Information Processing, Department of Electrical and Computer Engineering, Mississippi State University, Box 9571, 216 Simrall, Hardy Rd., Mississippi State, Mississippi 39762, 1998.

[384] L. R. Rabiner. A tutorial on Hidden Markov Models and selected applications in speech recognition. *Proceedings of the IEEE*, 77(2):257–286, February 1989.

[385] Kristie Seymore, Andrew Mccallum, and Ronald Rosenfeld. Learning hidden markov model structure for information extraction. 1999.

[386] Jose DeFigueiredo Coutinho, Matti Juvonen, Jun Wang, Benny Lo, Wayne Luk, Oskar Mencer, and Guang-Zhong Yang. Designing a Posture Analysis System with Hardware Implementation. *Journal of VLSI Signal Processing*, 2006.

[387] Howard Demuth and Mark Beale. *Neural Network Toolbox: For use with MATLAB: User's Guide, Version 4.* The Mathworks, Inc., 3 Apple Hill Drive, Natick, MA, USA, 2001.

[388] Manfred Morari and N. Lawrence Ricker. *Model Predictive Control Toolbox: For use with MATLAB: User's Guide, Version 1.* The Mathworks, Inc., 3 Apple Hill Drive, Natick, MA, USA, 1998

[389] *MATLAB: The Language of Technical Computing, Version 6.* The Mathworks, Inc., 3 Apple Hill Drive, Natick, MA, USA, 2002.

[390] George A. Miller. The Magical Number Seven, Plus or Minus Two: Some Limits on Our Capacity for Processing Information. *The Psychological Review*, 63:81–97, 1956. Online, last retrieved August 29, 2009. http://www.musanim.com/miller1956/.

[391] Alberto Gallace, Hong Z. Tan, and Charles Spence. The failure to detect tactile change: A tactile analogue of visual change blindness. *Psychonomic Bulletin 38; Review*, 13:300–303, April 2006.

[392] Alberto Gallace, Hong Z. Tan, and Charles Spence. Do "mudsplashes" induce tactile change blindness? *Perception 38; Psychophysics*, 69:477–486, May 2007.

[393] Alberto Gallace, Malika Auvray, Jessica Hartcher-O'Brien, Hong Z. Tan, and Charles Spence. The effect of visual, auditory, and tactile distractors on people's awareness of tactile change. In *Body Representation Workshop*, page 1, Rovereto, Italy, October 8–10, 2007.

[394] Vincent Hayward. A Brief Taxonomy of Tactile Illusions and Demonstrations That Can Be Done In a Hardware Store. *Brain Research Bulletin (Special Issue on Robotics and Neuroscience)*, 75(6):742–752, 2008.

[395] Hong Z. Tan. Haptic interfaces for virtual environments: perceived instability and force constancy in haptic sensing of virtual surfaces. *Canadian Journal of Experimental Psychology*, 61(3):265–275, September 2007.

[396] Hong Z. Tan, Mandayam A. Srinivasan, Charlotte M. Reed, and Nathaniel I. Durlach. Discrimination and identification of finger joint-angle position using active motion. *ACM Transactions on Applied Perception (TAP)*, 4(2):10, 2007. ISSN:1544-3558.

[397] Ethan D. Montag. Experimental Psychology, University of California, San Diego. Online, last retrieved August 29, 2009. http://www.cis.rit.edu/people/faculty/montag/vandplite/pages/chap_3/ch3p1.html.

[398] Harvey Richard Schiffman. *Sensation and perception: An integrated approach.* John Wiley & Sons, Inc., 1976. Rutgers, the State University, 434 pages, 1st edition, ISBN: 0-471-76091-9.

[399] Internet Sensation & Perception Laboratory University of South Dakota. Weber's Law of Just Noticeable Differences. Online, last retrieved July 14, 2008. http://www.usd.edu/psyc301/WebersLaw.htm.

[400] Unit of physiology Uppsala universitet, Department of neuroscience. Subjective vs perceived intensity. Online, last retrieved August 29, 2009. http://www.neuro.uu.se/fysiologi/gu/nbb/lectures/SubPerc.html.

[401] Wilfrid Jänig and Robert F. Schmidt. Single unit responses in the cervical sympathetic trunk upon somatic nerve stimulation. *Pflügers Archiv European Journal of Physiology*, 314(3):199–216, September 1970.

[402] G. A. Gescheider, S. J. Bolanowski, and R. T. Verrillo. Some characteristics of tactile channels. *Behavioural Brain Research*, 148:35–40 (6), January 5, 2004.

[403] Andreas Riener. Age- and Gender-Related Studies on Senses of Perception for Human-Vehicle-Interaction. In *2nd Workshop on Automotive User Interfaces and Interactive Applications (AUIIA'08), Mensch und Computer 2008, Luebeck, Germany*, volume 2. Logos Verlag, Berlin, September 7–10 2008. ISBN: 978-3-8325-2007-6, 8 pages.

[404] A. Jain, L. Hong, and S. Pankanti. Biometric identification. In *Communications of the ACM*, volume 43, pages 90–98, New York, NY, USA, 2000.

[405] Andreas Riener and Alois Ferscha. Supporting Implicit Human-to-Vehicle Interaction: Driver Identification from Sitting Postures. In *The First Annual International Symposium on Vehicular Computing Systems (ISVCS 2008)*, July 22-24, 2008, Trinity College Dublin, Ireland, July 2008. ACM Digital Library. 10 pages, ISBN: 978-963-9799-27-1.

[406] W. T. Phenice. A newly developed visual method of sexing the os pubis. *American Journal of Physical Anthropology*, 30:297–302, 1969.

[407] Eugene Giles. Discriminant function sexing of the human skeleton. *Personal Identification in Mass Disasters*, pages 99–109, 1970. Smithsonian Institution. Washington, D.C.

[408] T. D. Stewart. Sex determination of the skeleton by guess and by measurement. *American Journal of Physical Anthropology*, 12:385–392, 1954.

[409] Craig Schlenoff and Michael Gruninger. Towards a Formal Representation of Driving Behaviors. *Formal Approaches to Agent-Based Systems*, pages 290–291, 2002.

[410] Hongwei Qi, Qiangze Feng, Bangyong Liang, and Toshikazu Fukushima. Ontology-Based Mobile Information Service Platform. In Y. Zhang et al., editor, *Progress in WWW Research and Development*, number 4976 in Lecture Notes in Computer Science, pages 239–250, Springer-Verlag Berlin Heidelberg, 2008.

[411] Simone Fuchs, Stefan Rass, Bernhard Lamprecht, and Kyandoghere Kyamakya. A model for ontology-based scene description for context-aware driver assistance systems. In *Proceedings of the 1st international conference on Ambient media and systems (Ambi-Sys'08)*, pages 1–8, ICST, Brussels, Belgium, Belgium, 2008. ICST (Institute for Computer Sciences, Social-Informatics and Telecommunications Engineering).

[412] N. Baumgartner, W. Retschitzegger, and W. Schwinger. Lost in Space and Time and Meaning – An Ontology-Based Approach to Road Traffic Situation Awareness. In *Proceedings of the 3rd Workshop on Context Awareness for Proactive Systems (CAPS 2007)*, Guildford, United Kingdom, June 2007.

[413] Olivier Georgeon. Analyzing Traces of Activity for Modeling Cognitive Schemes of Operators. *The Society for the Study of Artificial Intelligence and Simulation of Behaviour (AISB), Quarterly Newsletter*, March 2008. 3 pages, Online, last retrieved August 31, 2009. http://www.aisb.org.uk/aisbq/qarchive.shtml.

[414] Anna Dubois. *Organising Industrial Activities Across Firm Boundaries*, chapter Activities, activity chains and activity structures, pages 25–37. Routledge Chapman & Hall, London – New York, 1998. ISBN: 978-0415147071.

[415] Peter R. Stopher, David T. Hartgen, and Yuanjun Li. SMART: Simulation model for activities, resources and travel. *Transportation*, 23(3):293–312, August 1996.

[416] Kay W. Axhausen, Andrea Zimmermann, Stefan Schönfelder, Guido Rindsfüser, and Thomas Haupt. Observing the rhythms of daily life: A six-week travel diary. *Transportation*, 29(2):95–124, May 2002.

[417] S. Maerivoet and B. De Moor. Transportation Planning and Traffic Flow Models. *ArXiv Physics e-prints*, July 2005.

[418] H. Hammadou, I. Thomas, D. van Hofstraeten, and A. Verhetsel. *Across the border – building upon a quarter century of transport research in the Benelux*, chapter Distance decay in activity chains analysis. A Belgian case study, pages 1–27. Uitgeverij De Boeck, 1st edition, November 2003. ISBN: 978-9-045-50956-3.

[419] Kari Torkkola, Mike Gardner, Chris Schreiner, Keshu Zhang, Bob Leivian, Harry Zhang, and John Summers. Understanding Driving Activity Using Ensemble Methods. In *Computational Intelligence in Automotive Applications*, pages 39–58. Springer-Verlag Berlin, Heidelberg, 2008. ISBN: 978-3-540-79256-7.

[420] Erwin R. Boer. Behavioral entropy as a measure of driving performance. In *International Driving Symposium on Human Factors in Driver Assessment, Training and Vehicle Design*, Snowmass Village at Aspen, Colorado, August 2001. 5 pages.

[421] Andreas Riener, Alois Ferscha, and Michael Matscheko. Intelligent Vehicle Handling: Steering and Body Postures while Cornering. In *21st International Conference on Architecture of Computing Systems (ARCS 2008), System Architecture and Adaptivity*, page 14. Springer-Verlag Berlin Heidelberg, February 2008.

[422] Jukka Linjama and Topi Kaaresoja. Novel, minimalist haptic gesture interaction for mobile devices. In *Proceedings of the third Nordic conference on Human-computer interaction (NordiCHI '04)*, pages 457–458, New York, NY, USA, 2004. ACM.

[423] Dinesh K. Pai. *Advances in Telerobotics*, volume 15 of *Springer Tracts in Advanced Robotics*, chapter Multisensory Interaction: Real and Virtual, pages 489–498. Springer Berlin Heidelberg, Computer Science Department, Rutgers University, Piscataway, NJ, August 2005. ISSN: 1610-7438 (Print) 1610-742X (Online).

[424] Eric H. Chudler. University of Washington Engineered Biomaterials (UWEB), Department of Bioengineering, Two-point Discrimination. Online, last retrieved August 26, 2009. http://faculty.washington.edu/chudler/twopt.html.

[425] Eric Gunther, Glorianna Davenport, and Sile O'Modhrain. Cutaneous grooves: composing for the sense of touch. In *NIME '02: Proceedings of the 2002 conference on New interfaces for musical expression*, pages 1–6, Singapore, Singapore, 2002. National University of Singapore.

[426] Andreas Riener and Alois Ferscha. Driver Activity Recognition from Sitting Postures. In Thilo Paul-Stueve, editor, *Workshop Automotive User Interfaces, hold in conjunction with 7th Mensch und Computer Conference, Weimar, Germany*. Institute for Pervasive

Computing, Johannes Kepler University Linz, Austria, Verlag der Bauhaus-Universität Weimar, September 2–5, 2007. 9 pages, ISBN-13: 978-3-486-58496-7.

[427] Andreas Riener and Alois Ferscha. Simulation Driven Experiment Control in Driver Assistance Assessment. In David Roberts, editor, *12th IEEE International Symposium on Distributed Simulation and Real Time Applications, Vancouver, BC, Canada*, number P3425. IEEE Computer Society, October 27–29, 2008. ISBN: 978-0-7695-3425-1, 10 pages.

[428] Andreas Riener and Alois Ferscha. Reconfiguration of Vibro-tactile Feedback based on Drivers' Sitting Attitude. In *Proceedings of the Second International Conference on Advances in Computer-Human Interactions (ACHI 2009), Cancun, Mexico*. IEEE Computer Society Press, February 1–7 2009.

[429] C. J. Golden and B. Schneider. Cell phone use and visual attention. *Perceptual and motor skills*, 97(2):9, October 2003.

[430] James McKnight and A. Scott McKnight. The Effect of Cellular Phone Use upon Driver Attention . *Ergonomics Abstracts, Accident Analysis & Prevention*, 25(3):259–265, 1991.

[431] Paul Green. Visual and Task Demands of Driver Information Systems. Report UMTRI 98-16, The University of Michigan, Transportation Research Institute (UMTRI), 2901 Baxter Rd, Ann Arbor, Michigan, June 1999. 119 pages.

[432] A. J. Brammer, J. E. Piercy, S. Nohara, H. Nakamura, and P. L. Auger. Age-related changes in mechanoreceptor-specific vibrotactile thresholds for normal hands. *The Journal of the Acoustical Society of America*, 93(4):p.2361, April 1993.

[433] Scott W Shaffer and Anne L Harrison. Aging of the Somatosensory System: A Translational Perspective. *Physical Therapy*, 87(2):193–207, February 2007.

[434] J. A. Smither, M. Mouloua, P. A. Hancock, J. Duley, R. Adams, and K. Latorella. *Human performance, situation awareness and automation: Current research and trends*, chapter Aging and Driving Part I: Implications of Perceptual and Physical Changes, pages 315–319. Erlbaum, Mahwah, NJ, 2004.

[435] Konrad W. Kallus, Jeroen A. J. Schmitt, and David Benton. Attention, psychomotor functions and age. *European Journal of Nutrition*, 44(8):465–484, December 2005. ISSN: 1436-6207 (Print) 1436-6215 (Online).

[436] Linda Breytspraak. Center on Aging Studies, University of Missouri-Kansas City. Online, last retrieved August 30, 2009. http://missourifamilies.org/quick/agingqa/agingqa18.htm.

[437] J. Adam, F. Paas, M Buekers, I. Wuyts, W. Spijkers, and P. Wallmeyer. Gender differences in choice reaction time: evidence for differential strategies. *Ergonomics*, 42(2):327–335, February 1999.

[438] S. Dane and A. Erzurumluoglu. Sex and handedness differences in eye-hand visual reaction times in handball players. *International Journal of Neuroscience*, 113:923–929(7), July 2003.

[439] Geoff Der and Ian J. Deary. Age and Sex Differences in Reaction Time in Adulthood: Results From the United Kingdom Health and Lifestyle Survey. *Psychology and Aging*, 21(1):62–73, 2006.

[440] Oliver Speck, Thomas Ernst, Jochen Braun, Christoph Koch, Eric Miller, and Linda Chang. Gender differences in the functional organization of the brain for working memory. In *Neuroreport*, volume 11, pages 2584–2585. Lippincott Williams & Wilkins, Inc., August 3, 2002.

[441] Jerome Barral and Bettina Debu. Aiming in adults: Sex and laterality effects. *Laterality: Asymmetries of Body, Brain and Cognition*, 9(3):299–312, 2004.

[442] C. J. Bellis. Reaction time and chronological age. *Proceedings of the Society for Experimental Biology and Medicine*, 30:801, 1933.

[443] B. Engel, P. Thorne, and R. Quilter. On the relationship among sex, age, response mode, cardiac cycle phase, breathing cycle phase, and simple reaction time. *Journal of Gerontology*, 27:456–460, 1972.

[444] Peter Bengtsson, Camilla Grane, and J. Isaksson. Haptic/graphic interface for in-vehicle comfort functions. In *Proceedings of the 2nd IEEE International Workshop on Haptic, Audio and Visual Environments and Their Applications (HAVE'2003)*, pages 25–29, Piscataway, 2003. IEEE Instrumentation and Measurement Society. ISBN: 0-7803-8108-4.

[445] Angelos Amditis, Aris Polychronopoulos, Luisa Andreone, and Evangelos Bekiaris. Communication and interaction strategies in automotive adaptive interfaces. *Cogn. Technol. Work*, 8(3):193–199, 2006.

[446] Vanessa Harrar and Laurence R. Harris. The effect of exposure to asynchronous audio, visual, and tactile stimulus combinations on the perception of simultaneity. *Experimental Brain Research*, 186(4):517–524, April 2008. ISSN: 0014-4819 (Print) 1432-1106 (Online).

[447] Chanon M. Jones, Rob Gray, Charles Spence, and Hong Z. Tan. Directing visual attention with spatially informative and spatially noninformative tactile cues. *Experimental Brain Research*, 186(4):659–669, April 2008. ISSN: 0014-4819 (Print) 1432-1106 (Online).

[448] Driver distraction trends and issues. *IEE Computing & Control Engineering Journal*, 16(1):28–30, February -March 2005.

[449] Andreas Riener and Alois Ferscha. Raising awareness about space via vibro-tactile notifications. In *3rd IEEE European Conference on Smart Sensing and Context (EuroSSC), Zurich, Switzerland*, Lecture Notes in Computer Science (LNCS). Springer Berlin / Heidelberg, October 29–31 2008. 12 pages.

[450] Nuria Oliver and Alex Pentland. Graphical models for driver behavior recognition in a SmartCar. In *Intelligent Vehicles Symposium, 2000. IV 2000. Proceedings of the IEEE*, pages 7–12, October 2000.

[451] Nuria Oliver and Alex Pentland. Driver behavior recognition and prediction in a SmartCar. In J. G. Verly, editor, *Society of Photo-Optical Instrumentation Engineers (SPIE) Conference Series*, volume 4023 of *Society of Photo-Optical Instrumentation Engineers (SPIE) Conference Series*, pages 280–290, June 2000.

[452] A. Aleksandrowicz and S. Leonhardt. Wireless and Non-contact ECG Measurement System - the "Aachen SmartChair". *Acta Polytechnica, Journal of Advanced Engineering*, 47(4–5):68–71 (4), 2007.

[453] Corinne Mattmann, Frank Clemens, and Gerhard Tröster. Sensor for Measuring Strain in Textile. *Sensors*, 8(6):3719–3732, June 2008.

[454] Brussels European Opinion Research Group. Survey on the behavior of people in the EU concerning physical activity. detailed data on time we spend sitting/day. *Eurobarometer - Physical Activity*, pages 8–15, 2003.

[455] Engineering Acoustics, Inc. (EAI). C-2 Tactor Datasheet. Online, last retrieved August 30, 2009. http://www.eaiinfo.com//PDFDocuments/C-2tactor.pdf.

[456] John D. Woodward, Christopher Horn, Julius Gatune, and Aryn Thomas. Biometrics – A Look at Facial Recognition. Technical report, RAND Public Safety and Justice,

RAND, 1700 Main Street, P.O. Box 2138, Santa Monica, CA 90407-2138, 2003. ISBN: 0-8330-3302-6.

[457] A. Bobick. Movement, activity and action: the role of knowledge in the perception of motion, 1997.

[458] P. Jonathon Philips, Alvin Martin, C. L. Wilson, and Mark Przybocki. An introduction to evaluating biometric systems. *IEEE Computer*, pages 56–63, February 2000. National Institute of Technology.

[459] Charles Scott Sherrington. The integrative action of the nervous system. 1906. New Heaven CT, Yale University Press.

[460] Anthony P. Scinicariello, Kenneth Eaton, Timothy J. Inglis, and J. J. Collins. Enhancing human balance control with galvanic vestibular stimulation. *Biological Cybernetics*, 84(6):475–480, May 2001.

[461] D. L. Wardman, J. L. Taylor, and R. C. Fitzpatrick. Effects of galvanic vestibular stimulation on human posture and perception while standing. *Journal of physiology*, 551(3):1033 – 1042 (10), September 2003.

[462] Joseph M. Furman, Martijn L. Müller, Mark S. Redfern, and J. Richard Jennings. Visual-vestibular stimulation interferes with information processing in young and older humans. *Experimental Brain Research*, 152(3):383–392, October 2003.

[463] Ann M. Bacsi and James G. Colebatch. Evidence for reflex and perceptual vestibular contributions to postural control. *Experimental Brain Research*, 160(1):22–28, January 2005.

[464] Hamish MacDougall, Steven Moore, Ian Curthoys, and F. Black. Modeling postural instability with galvanic vestibular stimulation. *Experimental Brain Research*, 172(2):208–220, June 2006.

[465] Cecilia Laschi, Eliseo Maini, Francesco Patane, Luca Ascari, Gaetano Ciaravella, Ulisse Bertocchi, Cesare Stefanini, Paolo Dario, and Alain Berthoz. A Vestibular Interface for Natural Control of Steering in the Locomotion of Robotic Artifacts: Preliminary Experiments. *Robotics Research*, pages 537–551, May 2007.

[466] T. Maeda, H. Ando, T. Amemiya, N. Nagaya, M. Sugimoto, and M. Inami. Shaking the world: galvanic vestibular stimulation as a novel sensation interface. In *International Conference on Computer Graphics and Interactive Techniques, ACM SIGGRAPH 2006 Emerging Technologies, Los Angeles, CA, USA*, New York, NY, USA, 2005. ACM.

[467] Wolfgang Taube, Christian Leukel, and Albert Gollhofer. Influence of enhanced visual feedback on postural control and spinal reflex modulation during stance. *Experimental Brain Research*, 188(3):353–361, July 2008.

[468] Allen Sandler and Susan McLain. Use of Noncontingent Tactile and Vestibular Stimulation in the Treatment of Self-injury: An Interdisciplinary Study. *Journal of Developmental and Physical Disabilities*, 19(6):543–555, December 2007.

[469] Tomohiro Amemiya, Hideyuki Ando, and Taro Maeda. Perceptual attraction force: exploit the nonlinearity of human haptic perception. In *International Conference on Computer Graphics and Interactive Techniques, ACM SIGGRAPH 2006 Sketches, Los Angeles, CA, USA*, New York, NY, USA, 2006. ACM.

[470] Hirofumi Aoki, Charles M. Oman, Alan Natapoff, and Andrew Liu. The effect of the configuration, frame of reference, and spatial ability on spatial orientation during virtual 3-dimentional navigation training. In *Seventh Symposium on the Role of Vestibular Organs in Space Exploration (ESTEC'06), Noordwijk, The Netherlands*, June 2006. 2 pages.

[471] Laurence R. Harris, Richard Dyde, Charles M. Oman, and Michael Jenkin. Visual cues to the direction of the floor. In *Seventh Symposium on the Role of Vestibular Organs in Space Exploration (ESTEC'06), Noordwijk, The Netherlands*, June 2006. 2 pages.

[472] Naohisa Nagaya, Maki Sugimoto, Hideaki Nii, Michiteru Kitazaki, and Masahiko In ami. Visual perception modulated by galvanic vestibular stimulation. In *Proceedings of the 2005 international conference on augmented tele-existence (ICAT'05)*, pages 78–84, New York, NY, USA, 2005. ACM.

[473] Heather L. Jenkin, James E. Zacher, Michael R. Jenkin, Charles M. Oman, and Laurence R. Harris. Effect of Field of View on Visual Reorientation Illusion: Does the levitation illusion depend on the view seen or the scene viewed? In *Seventh Symposium on the Role of Vestibular Organs in Space Exploration (ESTEC'06), Noordwijk, The Netherlands*, June 2006. 2 pages.

[474] John F. Golding, Priscilla Kadzere, and Michael A. Gresty. Motion Sickness Susceptibility Fluctuates Through the Menstrual Cycle. *Aviation, Space, and Environmental Medicine*, 76:970–973(4), October 2005.

[475] Christopher E. Carr and Dava J. Newman. When is Running More Efficient Than Walking in a Space Suit? In *35th International Conference on Environmental Systems (ICES),*

Rome, Italy, number 2005-01-2970 in SAE Technical Papers Series, Warrendale, PA, USA, July 2005. SAE International. ISSN: 0148-7191, 6 pages.

[476] Jukka Raisamo. Haptic User Interfaces: The Sense of Touch. Multimodal Interaction Group, Tampere Unit for Computer-Human Interaction, Department of Computer Sciences. University of Tampere, Finland, Online, last retrieved August 29, 2009. http://www.cs.uta.fi/hui/lectures/, 2007.

[477] Gabriel Robles-De-La-Torre. Re: FW: Haptics, Tactile, Proprioception. E-Mail, received Wed, 10 Dec 2008 15:36:01 -0600, December 10, 2008. Gabriel@RoblesDeLaTorre.com, International Society for Haptics.

[478] F. J. Clark, R. C. Burgess, and J. W. Chapin. Proprioception with the Proximal Interphalangeal Joint of the Index Finger: Evidence for a Movement Sense Without a Static-Position Sense. *Brain, A Journal of Neurology*, 109(6):1195–1208, 1986.

[479] Susan J. Lederman. Re: Haptics, Tactile, Proprioception. E-Mail, received Wed, 10 Dec 2008 07:31:39 -0500, December 10, 2008. susan.lederman@queensu.ca, Department of Psychology, Queen's University, Kingston, Ontario, Canada K7L 3N6.

[480] Roberta L. Klatzky. Re: Haptics, Tactile, Proprioception. E-Mail, received Wed, 10 Dec 2008 09:53:35 -0500, December 10, 2008. klatzky@cmu.edu, Department of Psychology, Carnegie Mellon University, Pittsburgh, Pennsylvania 15213.

[481] Jeremy M. Wolfe, Keith R. Kluender, Dennis M. Levi, Linda M. Bartoshuk, Rachel S. Herz, Roberta L. Klatzky, Susan J. Lederman, and Daniel M. Merfeld. *Sensation & Perception, Second Edition (Hardcover)*. Sinauer Associates Incorporated, 2008. ISBN-13: 978-087893-953-4, 450 pages.

[482] Vincent Hayward. Re: Haptics, Tactile, Proprioception. E-Mail, received Tue, 09 Dec 2008 23:18:30 +0100, December 9, 2008. hayward@cim.mcgill.ca, Institut des Systemes Intelligents et de Robotique, Université Pierre et Marie Curie, Paris, France.

[483] Hong Z. Tan. Re: Haptics, Tactile, Proprioception. E-Mail, received Thu, 25 Dec 2008 14:59:34 -0500, December 25, 2008. hongtan@purdue.edu, Purdue University, Electrical Engineering Building, 465 Northwestern Avenue, West Lafayette, Indiana 47907-2035.

[484] F. J. Clark and K. W. Horch. *Handbook of Perception and Human Performance: Sensory Processes and Perception*, volume 1, chapter Kinesthesia, pages 13/1 – 13/62. New York: Wiley, 1986.

[485] J. M. Loomis and S. J. Lederman. *Handbook of Perception and Human Performance: Cognitive Processes and Performance*, volume 2, chapter Tactual Perception, pages 31/1–31/41. New York: Wiley, 1986.

[486] Henry Piéron. Géza Révész: 1878-1955. *The American Journal of Psychology*, 69(1):139–141, March 1956. Published by: University of Illinois Press.

[487] Vittorio Busato. 100 jaar Psychologie aan de Universiteit van Amsterdam: Korte biografie van Géza Révész (2006). Online, last retrieved December 12, 2008. http://www.100jaarpsychologie.uva.nl/gesch1.php.

[488] Edwin G. Boring. *Sensation and Perception in the History of Experimental Psychology*. Appleton Century Crofts, New York, 1942. 644 pages.

[489] James J. Gibson. *Die Sinne und der Prozess der Wahrnehmung (Gebundene Ausgabe)*. Verlag Hans Huber, Bern, 1973. 397 Seiten, 1. Auflage, ISBN: 3-456-30586-9, Originaltitel: The Senses considered as Perceptual Systems, 1966, Houghton Mifflin Company.

[490] Wikipedia. Proprioception. Online, last retrieved August 30, 2009. http://en.wikipedia.org/wiki/Proprioception.

[491] R. J. van Beers, Anne C. Sittig, and Jan J. Denier van der Gon. The precision of proprioceptive position sense. *Experimental Brain Research*, 122(4):367–377, October 1998.

[492] Lesley A. Hall and D. I. McCloskey. Detections of movements imposed on finger, elbow and shoulder joints. *The Journal of Physiology*, 335:519–533, February 1983.

[493] Franziska K.B. Freyberger, Martin Kuschel, Berthold Farber, Martin Buss, and Roberta L. Klatzky. Tilt Perception by Constant Tactile and Constant Proprioceptive Feedback Through a Human System Interface. *EuroHaptics Conference, 2007 and Symposium on Haptic Interfaces for Virtual Environment and Teleoperator Systems. World Haptics 2007. Second Joint*, pages 342–347, March 2007.

[494] Julien Voisin, Yves Lamarre, and Elaine Chapman. Haptic discrimination of object shape in humans: contribution of cutaneous and proprioceptive inputs. *Experimental Brain Research*, 145(2):251–260, July 2002.

[495] O. E. Surnina and E. V. Lebedeva. Sex- and Age-Related Differences in the Time of Reaction to Moving Object in Children and Adults. *Human Physiology*, 27(4):436–440, July 2001.

[496] Robert J. Kosinski. A Literature Review on Reaction Time(September 2008). Clemson University, Department of Biological Sciences, last retrieved January 15, 2009, http://biology.clemson.edu/bpc/bp/Lab/110/reaction.htm.

[497] F. Galton. On instruments for (1) testing perception of differences of tint and for (2) determining reaction time. *Journal of the Anthropological Institute*, 19:27–29, 1899.

[498] K. von Fieandt, A. Huhtala, P. Kullberg, and K. Saarl. Personal tempo and phenomenal time at different age levels. Reports from the Psychological Institute No. 2, University of Helsinki, 1956.

[499] A.T. Welford. *Reaction time*, chapter Choice reaction time: Basic concepts, pages 73–128. Academic Press, New York, 1980. ISBN: 0127428801.

[500] Alan Dix. Closing the Loop: Modelling action, perception and information. In T. Catarci, M. F. Costabile, S. Levialdi, and G. Santucci, editors, *AVI'96 - Advanced Visual Interfaces*, pages 20–28. ACM Press, 1996.

[501] John. A. Hoxmeier and Chris DiCesare. System Response Time and User Satisfaction: An Experimental Study of Browser-based Applications. In *Proceedings of the Fourth CollECTeR Conference on Electronic Commerce*, Breckenridge, Colorado, USA, April 2000. Collaborative Electronic Commerce Technology and Research (CollECTeR).

[502] Thomas J. Triggs and Walter G. Harris. Reaction Times of Drivers to Road Stimuli. Human Factors Report HFR-12, Human Factors Group, Monash University (Accident Research Centre), Department of Psychology, Monash University, Victoria 3800, Australia, June 1982. ISBN: 0-86746-147-0.

[503] Leah D. Sheppard and Philip A. Vernon. Intelligence and speed of information-processing: A review of 50 years of research. *Personality and Individual Differences*, 44(3):535–551 (17), 2008.

[504] M. M. Wickremaratchi and J. G. Llewelyn. Effects of ageing on touch. *Postgraduate Medical Journal*, 82(967):301–304, May 2006.

[505] Wikipedia. Fingerspelling. Online, last retrieved August 31, 2009. http://en.wikipedia.org/wiki/Fingerspelling.

Index

Activity Recognition
  from Sitting Postures, 132
  while Cornering, 138
ADAS, IX, 3, 9, 35, 75, 83, 148, 229
  Overview, 83
Age Sensitivity, 117
Allen Relationships, 72
Alphabets, 230
  Auditory, 230
  Haptic, 93, 230
    Optohapt, 232
    Vibratese Language, 79, 231
  Visual, 230
American Sign Language, 230
Amodal, 24
Analytical Methods, 107
Appendices, 197
Application Domains, 29
Articulation, 27
  Application Domains, 27

Biometric
  Features, Including of, 195
  Identification, 207
Biometric Identification, 207
  Authentication, 207
  Definition, Terms of, 207
  Ideal Systems, 211
  Persons in Vehicles, 208
  The Process, 210
    Closed Set, 210
    False Accept Rate, 211
    False Reject Rate, 211
    Open Set, 210
  Verification, 207
Biosignal Sensing, 204
Body Postures, Experiments on, 132, 138
Braille, 230

CAN bus, 3, 27, 80, 206
Car Seat
  Vibro-tactile Interaction, 61
Cognitive Load
  Challenges, 67
  Definition, 68
  Empirical Evidence, 68
  Managing Workload, 69
    Examples, 70
Cognitive Overload
  from Interaction with ADAS, 162
  from Physical Limitations, 162
Conclusions, 183, 194
Cutaneous, Definition of Terms, 48

Dependency, Deterioration
  Age, 162
  Gender (Sex), 163
Discussion, 183
Distraction
  Classes of, 162
  Factors, 8
  Forecast, 12

Types of, 11
Distraction, Causes of, 73
Driver Activity
Driver Distraction, 72
How, 76
What Information?, 74
When, 76
Driver Assistance Assessment, 161
Driver Assistance Systems, IX, 30, 41, 83, 119, 138, 205
Driver Demands
Activity, 71
In-car Perception, 80
Notification, 71
Driver Information, Categorization of, 75
Capacity Limits, 77
Driver-Vehicle Expression, 18
Driver-Vehicle Interaction, 4, 39, 67
Driving
Risks of, 83

ECG, 204
ECU, 3, 96, 206
Electrocardiogram (ECG), 84, 109, 141, 204, 213
electrocardiogram (ECG), 195, 204
Experiments, 119
Driver Articulation
Activity Recognition, 132, 138
Identification, Authentication, 121
Driver-Vehicle Input, 120
Haptic Perception, 147
Dynamic Vibro-tactile Feedback, 148
Further Experiments, 182
Simulation-Driven Design, 161
Explicit Interaction, 19

Fingerspelling (ASL), 230
FlexRay, 3, 27, 80, 96, 206
Force Displays, 49
Future Prospects, 195

Galvanic Skin Response (GSR), 131, 205, 213
Gender Sensitivity, 117
GPS, 52, 98, 136, 205, 216

Haptic, 221
Haptic Processing, 148
Haptic Stimulation
Age Dependency, 227
Gender Difference, 228
Haptics, Definition of Terms, 48
Haptics, Novelty of, 192
Hardware, 199
Pressure Sensing, 199
Vibration Transmission, 202
Human Behavior, 207
Human Factors
Age Dependency, 225, 227
Gender Dependency, 225
Gender Difference, 228
Reaction Time, 133, 225
Human-Computer Interaction (HCI), 3, 20, 27, 41, 171, 190
Hypotheses, 30
Reflecting on the, 186

I/O Comparison Table, 29
Identification
Biometric, 207
Potential, 213
False Accept Rate (FAR), 211
False Reject Rate (FRR), 211
Limitations
Biometric, 213
Implicit Interaction, 20

In-car Auditory Perception, 80
In-car Interaction, State of the Art, 41
  Gustation, Taste, 45
  Hearing, 42
  Olfaction, Smell, 43
  Vision, 42
In-car Visual Perception, 80
Information
  Notification Level, 76
Interaction in Vehicles, 162
Interaction Loop, Driver-Vehicle
  Activities, 67
  Information Demands, 67
Interaction Modalities, 29
Interaction, Modes of
  Amodal, 24
  Explicit, 19
  Implicit, 20
  Multimodal, 21
  Unimodal, 21
Interface Paradigms, 6
Interfaces
  Vibro-tactile, 85
Introduction, 3
Investigation, Predeterminations for
  Domain, 185
  Prototype, 185
IVIS, IX, 8, 83

Just Noticeable Difference (JND), 107, 114

Kinesthesis, 222
  Definition of Terms, 48

Language
  Vibro-tactile, 93
Level of Attention (LOA), 17, 71, 100, 101, 103, 147, 196
LIN bus, 3, 27

Mechanoreceptors, Cutaneous
  Meissner Corpuscles, 91
  Merkel Receptors, 91
  Pacinian Corpuscles, 51, 91, 169, 223
  Ruffini Corpuscles, 90
Meissner Corpuscles, 91
Merkel Receptors, 91
Message Length, Auditory, 78
Methodology, 105
  Hidden Markov Models (HMM), 111
  Linear Discriminant Analysis (LDA), 111
  Multivariate Data Analysis (MDA), 110
  Posture Pattern Analysis, 109
  Pressure Sensing, 109
  Principal Component Analysis (PCA), 110
  Requirements, 107
  Technological Conditions, 107
  Vibro-tactile Stimuli Detection, 110
Miscommunication
  Preventing of, 7
Morse Code, 230
MOST bus, 3, 27, 96, 206
Multimodal Interfaces, 195
Multimodality, 7, 21, 79
  Reaction Times, 195
Multiple Resource Theory, 22, 45, 52, 81

Navigation
  Vestibular provoked, 215

OBD, 136, 140, 145, 206
OLE, 200
Optohapt, Optophone, 232
Output
  Dynamic Adaptation, 148

Pacinian Corpuscles, 50, 51, 89, 91, 116, 169, 203, 223
Park Distance Control, 15, 42
Perception, 27
  Application Domains, 27
  Environmental, 14
Perception, Senses of, 14
Pressure Mat Systems
  Tekscan, 202
  Vista Medical FSA, 200
  XSensor, 201
Pressure Sensing, 199
  Methods, 110
Proprioception, 217, 222
Proprioception, Definition of Terms, 48

Reaction Time, 195
Research Questions, 30
Respiration Sensor, 84
Response, Reaction Time, 225
  Alternatives, 226
Ruffini Corpuscles, 90

Sensing
  Biosignals, 204
  Environmental, 205
  Vibrations, 91
Sensor
  Vehicle-specific data, 206
Sensory Channels
  Physical Limitations, 162
Sensory Modalities
  Additional, 196
  Hearing, 15
  Kinesthetic, 217
  Movement, 217
  Overview, 41
  Physiological, 214
    Balance, 215
    Equilibrioception, 215
    Nociception, 217
    Pain, 217
    Temperature, 216
    Thermoception, 216
    Vestibular, 215
  Position, 217
  Proprioception, 214, 217
  Selection, 77
  Tactition, 46
  Touch, X, 16, 36, 46, 60, 79, 86, 93
    Definition of Terms, 48
  Vision, 15
    Design Guidelines, 78
Simulated Driving, Studies on, 161
Simulation
  Reasons, 161
Skin, 50
  Mechanoreceptors, Cutaneous, 89
  Stimulation Parameters, 89
Steven's Formula, 115
Stimulation
  Skin, 89
  Vibro-tactile, 89
Stimulation Technologies, 51
  Vibro-tactile, in a Car Seat, 61
  Vibro-tactile, in Vehicles, 58
  Vibro-Tactile, Related Work, 51
  Vibro-tactile, Related Work
    Arms, 58
    Bottom, Back, 53
    Buttocks, 57
    Feet, 55
    Hands, 57
    Head, 55
    Torso, 52

Stimulus Threshold, 116
System Design, 109

Tactile Display, Motivation for, 46
Tactile, Definition of Terms, 48
Tactogram, 93, 101, 167
Tactograms, 98, 196
    6-Tuple Notation, 100
    Dynamic View, 100
    Multi-Tactor Systems, 99
Tactors, C-2, 203
Tadoma, 231
The Weber-Fechner law, 114
Touch-Related Alphabets, 230

Unimodal, 21
User Interfaces, 3

Vibratese Alphabet, 231
Vibration Feedback, 202
Vibration Noise, 192
Vibration Perception Threshold, 94
Vibro-tactile Actuators, 202
    Mass Motors, 203
    Voice Coils, 203
Vibro-tactile Articulation, 41
Vibro-tactile Feedback, 148
    Reconfiguration, 195
Vibro-tactile Interfaces, 85
    Types of Stimuli, 87
        Electro-tactile, 87
        Heat, 88
        Mechanical, 88
        Thermal, 88
Vibro-tactile Pattern, 93
Vibro-tactile Patterns, 98
Vibro-tactile Presentation, 41
Vibro-tactile Stimulation, 95
Vibro-tactile Stimulation, Parameters of
    Activation Time, 96
    Distance, 95
    Frequency, 97, 116
    Response Time, 96
    Vibration Amplitude (Intensity), 97
Vibro-tactile Stimuli Detection
    Age and Gender, Dependency on, 117
    Discriminating Stimuli, 114
    Just Noticeable Difference (JND), 114
    Steven's Formula, 115
    Stimulus Threshold, 116
    The Weber-Fechner law, 114
    Weber's Law, 114
Vital Context
    In cars, 84

Where, 80

GPSR Compliance

The European Union's (EU) General Product Safety Regulation (GPSR) is a set of rules that requires consumer products to be safe and our obligations to ensure this.

If you have any concerns about our products, you can contact us on ProductSafety@springernature.com

In case Publisher is established outside the EU, the EU authorized representative is:

Springer Nature Customer Service Center GmbH
Europaplatz 3
69115 Heidelberg, Germany

Zeitfracht Medien GmbH
Ferdinand-Jühlke-Straße 7
99095 Erfurt, Deutschland
produktsicherheit@kolibri360.de